"Wheeler's compelling narrative reminds us of [illegible] sity and forcefully argues that the way forward, [illegible] humanity, must include and emphasize detailed m[illegible] species we name. This book is in part a provocative, fact-based opinion piece, a memoir of a lifelong passion for the wonders of the natural world, and a serious logical challenge to the hegemony of experimentalist and molecular genetics in biology. Wheeler's take is more than just a screed on the current state of affairs, he lays out a vision of the solution. His solution requires a reconstituted science of taxonomy, a collaborative global workforce, and funds to make it happen."

Kipling W. Will, *University of California, Berkeley*

"The author has succeeded in writing a highly original book on species exploration: their discovery, explanation, and relationships, that is both rigorous and accessible to a wide audience. There is no book on the market that addresses the nature of the content of this book with the breadth, depth, and clarity that this book achieves."

Antonio G. Valdecasas, *Museo Nacional de Ciencias Naturales, Madrid*

"Usually thought of as the science dealing with describing and classifying all organisms, taxonomy is a rich and varied discipline. Quentin Wheeler's book sets out to make a compelling case for taxonomy as a significant, fundamental, if under-appreciated, discipline. Rather than detail all the intricacies of its intellectual complexity, his book is a straightforward, no nonsense pragmatic account – one that should be, must be – read. But not just by life scientists and environmental policy makers, but by the general public as well – if for no other reason than to appreciate just how the interrelatedness of life on our earth is understood."

David M. Williams, *The Natural History Museum, London*

"*Species, Science and Society* is a constructive defense and promotion of Systematics in the 21st century... A timely book to remind us that good research in Systematics must integrate identification, description and classification in numerous and complementary comparative approaches... A clear statement that the results of Systematics are essential for a better understanding of Biodiversity, a fundamental societal challenge in facing global changes."

Thierry Bourgoin, *Museum National d'Histoire Naturelle, Paris*

"This excellent book is lighting a path for those who wish to respond to the biodiversity crisis with expanded taxonomic knowledge rather than retracted expectations. It makes a persuasive argument for a mainly scientific solution to the biodiversity crisis based on the science of Systematics to replace the poverty of contemporary frameworks that treat the natural world as ecosystem services, natural capitol and nature-based solutions which, based on current data, have failed to halt or address the decline of biodiversity in any significant way. At its heart this book extols the idea that we need to live with, understand and document the natural world rather than solely viewing it as an object for exploitation."

Robert Scotland, *Oxford University*

"A plea for a renaissance of taxonomy in its full form by one of the greatest living advocates of the field. Wheeler speaks to the next generation of researchers in a personal and often humorous narrative, warning that the foundation of biodiversity studies is rapidly crumbling."

Joseph V. McHugh, *University of Georgia*

SPECIES, SCIENCE AND SOCIETY

This book presents an engaging and accessible examination of the role of systematic biology in species exploration and biodiversity conservation.

Our planet and systematic biology are at a crossroads. Millions of species face an imminent threat of extinction, and, with knowledge of only a fraction of earth's species we are unprepared to respond. *Species, Science and Society* explains what is at stake if we continue to ignore the traditional mission of systematics. Rejecting claims that it is too late to document earth's species, that molecular evidence is sufficient and that comparative morphology and the grand traditions of systematics are outdated, this book makes a compelling argument for a taxonomic renaissance. The book challenges readers to rethink assumptions about systematics. Shattering myths and misconceptions and clarifying the role of systematics in confronting mass extinction, it hopes to inspire a new generation of systematists. Readers are given a deeply personal view of the mission, motivations and rewards of systematic biology. Written in narrative style with passion, wit and optimism, it is the first book to question the growing dominance of molecular data, defend descriptive taxonomy and propose a mission to discover, describe and classify all species. Our evolutionary heritage, the fate of society and the future of the planet depend on what we do next.

This book will be of great interest to academics, researchers and professionals working in systematics, taxonomy and biodiversity conservation, as well as students with a basic background in biology.

Quentin Wheeler was President of the State University of New York College of Environmental Science and Forestry, Vice President and Dean of the College of Liberal Arts and Sciences in Arizona State University, Keeper and Head of Entomology in the Natural History Museum, London, Director of the Division of Environmental Biology at the National Science Foundation, and Professor of Insect Systematics in Cornell University, U.S.A. He produces a weekly podcast and newsletter, The Species Hall of Fame, and his previous books include *The Future of Phylogenetic Systematics* (2016), *The New Taxonomy* (2008), *Letters to Linnaeus* (2009), *What on Earth?—100 of Our Planet's Most Amazing New Species* (2013), *Species Concepts and Phylogenetic Theory—A Debate* (2000), *Extinction and Phylogeny* (1992) and *Fungus-Insect Relationships* (1984).

Routledge Studies in Conservation and the Environment

This series includes a wide range of inter-disciplinary approaches to conservation and the environment, integrating perspectives from both social and natural sciences. Topics include, but are not limited to, development, environmental policy and politics, ecosystem change, natural resources (including land, water, oceans and forests), security, wildlife, protected areas, tourism, human-wildlife conflict, agriculture, economics, law and climate change.

Religion and Nature Conservation
Global Case Studies
Edited by Radhika Borde, Alison A. Ormsby, Stephen M. Awoyemi, and Andrew G. Gosler

Jackals, Golden Wolves, and Honey Badgers
Cunning, Courage, and Conflict with Humans
Keith Somerville

Case Studies of Wildlife Ecology and Conservation in India
Edited by Orus Ilyas and Afifullah Khan

Creating Resilient Landscapes in an Era of Climate Change
Global Case Studies and Real-World Solutions
Edited by Amin Rastandeh and Meghann Jarchow

Species, Science and Society
The Role of Systematic Biology
Quentin Wheeler

For more information about this series, please visit: www.routledge.com/Routledge-Studies-in-Conservation-and-the-Environment/book-series/RSICE

Species, Science and Society
The Role of Systematic Biology

Quentin Wheeler

LONDON AND NEW YORK

Designed cover image: © Getty Images

First published 2024
by Routledge
4 Park Square, Milton Park, Abingdon, Oxon OX14 4RN

and by Routledge
605 Third Avenue, New York, NY 10158

Routledge is an imprint of the Taylor & Francis Group, an informa business

British Library Cataloguing-in-Publication Data
A catalogue record for this book is available from the British Library

ISBN: 978-1-032-48439-6 (hbk)
ISBN: 978-1-032-48052-7 (pbk)
ISBN: 978-1-003-38907-1 (ebk)

DOI: 10.4324/9781003389071

Typeset in Goudy
by Newgen Publishing UK

For Marie

Contents

Preface *xi*
Acknowledgments *xviii*

PART I
Overview **1**

1 A Little about Molecules 3

2 Scientific Malpractice 8

3 The Science of Species 14

4 The Art of Survival 22

5 Cosmology of the Life Sciences 26

6 Choices 34

7 Everything You Always Wanted to Know about Taxonomy but Were Afraid to Ask 44

8 A Science Misunderstood Greatly 61

9 The Species-Scape 74

10 The Illusion of Knowledge 82

11 Morphology without Apology 92

12 The Inventory Imperative 104

13 Other than That, Mrs. Lincoln, How Was the Play? 118

PART II
A Crisis of Crises **127**

14 Extinction 129

15 Systematics under Siege 139

16 The Nature Gap 152

17 Options for a Sustainable Future 160

PART III
Solutions **165**

18 Taxonomic Renaissance 167

19 A Planetary-Scale Species Inventory 181

20 Hall of the Holocene 198

21 Shameless Self-Promotion 214

22 The Evolution of Evolutionary Economics 220

Epilogue 225
Glossary 233
Index 236

Preface

The aims of this book are to reassert the identity and mission of systematic biology, the science that explores and classifies species; to examine the benefits of doing so for the environment, conservation and understanding evolutionary history; and to ask what kind of future we want for science, society and the planet—and the role of systematics in creating it.

Systematic biology is changing as rapidly as the biosphere. It is important that basic aspects of systematics don't go extinct along with the plants, animals and microbes that it studies. As the rate of extinction increases, it is clear that the opportunity to fully explore species is disappearing. Abandoning the traditional goals of systematics now is not a choice that should be made lightly for it comes with profound and irreversible consequences.

This book assumes a basic background in biology. Written primarily for upper division undergraduate students interested in systematics, biodiversity, evolution, conservation and the environment, it raises issues that should concern everyone. It may be assigned as a supplemental class reading, used as the focus of a discussion group or read by individuals concerned by species extinction. Saving the traditional mission of systematics will require all hands on deck. Many systematists will welcome a defense of descriptive taxonomy. Others who have bought into the idea of a molecular-based taxonomy, or who see diminished value in natural history collections, will be challenged to rethink their positions.

A debate about the future of systematic biology is urgently needed. Believing that the pendulum has swung too far in the direction of a molecular-based taxonomy, I give it a hard shove in the opposite direction. My rhetoric will anger some, to which I say "Good!" I've been angered for years by the disrespect shown for the traditions and accomplishments of systematics. This book will succeed if readers contemplate the appropriate role of systematics in our response to mass extinction and of molecular data in systematics. I have attempted to stay out of the weeds with respect to highly technical issues, such as details of data analysis. These are contentious, highly consequential questions that I trust will be sorted out by systematists who reengage Hennig's theoretical revolution. My concern here is more fundamental, having to do with what systematic biology is and what it aims to achieve. Unless the battle to save systematics as an independent science is won, more technical questions about how it is done will matter little.

Although ideas in this book apply to systematics generally, my focus is on plants and animals. Microbes are a special case. Exactly what a microbial species is remains controversial. Because few microbes can be cultured in the laboratory where they can be studied in depth, microbiology is more dependent on molecular data than botany or zoology. As a result, microbial projects are more readily funded than descriptive studies of plants and animals. Microbes go extinct, too, of course. But recent evidence for bacteria, such as a study by Louca et al., suggests that they do not suffer similarly high losses during mass extinctions, making their inventory a little less urgent.

This is not a textbook, how-to guide or essay dense with citations. Purposefully written in narrative style, it shares something of the intellectual breadth and richness of traditional systematics. Hints in the text, such as author names, point the reader to references. Technical terms are defined in a glossary. Further readings back up statements in the text, provide examples or expand on ideas and are only intended as an entrée into the literature. Three books, in particular, are recommended as next readings regarding the extinction crisis, conservation and biomimicry: Kolbert, Wilson and Benyus, respectively. Good systematics textbooks include Williams & Ebach, Brower & Schuh and Judd et al.

On a slippery slope, sliding into a mass extinction, faced with the prospect of losing millions of species, we make our last stand for biodiversity now. What becomes of life on earth and humankind depends very much on what we do next. Responding to this crisis requires basic knowledge about species; knowledge that we do not have; knowledge that we have no plan to acquire. Worse, the science that creates such knowledge—systematic biology, known also as taxonomy or systematics—is misunderstood, maligned and marginalized.

This book is about what is happening to our planet's species and the science that studies them—and what we can do about it. It clarifies misconceptions about systematic biology, reimagining it for the challenges we face. It proposes an ambitious, planetary-scale project to inventory every kind of plant and animal on earth within decades. And, it urges natural history museums to once again be leaders of species exploration and taxonomic research. Nothing can better prepare us to avoid losses of biodiversity. Nothing of greater value can be gifted to future generations—condemned to life in a degraded biosphere—than knowledge of species.

Each year, tens of thousands of species go extinct reducing the number and diversity of living things, eliminating evidence of evolutionary history and removing players from the complex dynamics of the biosphere. We don't know exactly which or how many kinds of plants and animals are disappearing, but how could we? To date, we have discovered and named fewer than 20% of them. As we face the greatest environmental crisis in human history, we do so without the simplest facts.

We foolishly settle for superficial knowledge at a time when deep understanding is called for. Darwin convinced us that evolution took place. But in spite of impressive progress studying its mechanisms, from mutations to natural selection, we have only begun to piece together the actual history of evolution, the series of transformations that explains the diversity we see today. Ecologists have elucidated functions of ecosystems in spite of fractional awareness of the kinds of organisms present. We must question the sincerity of our commitment to biodiversity conservation so long as we fail to describe and name the species we profess to care about. An account of species should be among the first, and bare minimum, responses to extinction and ecosystem degradation. After decades of neglect, the dangers of disregarding such elementary knowledge are becoming apparent. We face planet-altering, irreversible decisions of enormous consequence in shameful, self-imposed ignorance.

Although we cannot say what species exist, there is no doubt about what is happening to them. From on-the-ground observations to orbiting satellites, every indicator tells the same story. As environmental impacts of human activities advance, the diversity of life retreats. As habitats are altered, fragmented and degraded, species are lost. It's that simple. Unless we do something to change the path we are on, 70% of the world's species will be gone in less than 300 years.

Aware of the accelerated loss of species, we can no longer justify inflicting savage scars on the biosphere or turning a blind eye to their fate. That most extinctions are invisible to us in no way reduces our culpability or the threat they represent to the biosphere and our future. Since the origin of life, about 3.8 billion years ago, there have been five comparable "mass extinction" events. These are periods in the geologic record when 70% or more of the world's species disappear. The most recent was tens of millions of years ago, long before humans existed. The extinction event now underway threatens to be similar in scale but differs in one important respect. It is driven by us. From deforestation to resource exploitation, land conversion and elevated atmospheric carbon dioxide, insatiable demands of a growing human population are devastating the natural world. More worrisome than the number of species lost is the random reduction of diversity. It may ultimately matter more which species survive than how many, although the more, the better. Unaware of species that exist, their status, habits and geography, we are alarmingly unprepared to make informed decisions as we enter an uncertain future.

We know far less about life on earth than most people suppose. Scientists who explore and classify the kinds of plants, animals and microbes have named about 2 million species since the beginning of "modern" taxonomy in 1758. A conservative estimate is that there are 10 million living species, although the actual number could be very much higher. Either way, we lack the information needed to take confident steps to mitigate losses.

It is difficult to say just how catastrophic a mass extinction could be allowed to proceed unabated. In the extreme, it could be an existential threat to civilization as we know it. Robbed of resources to meet basic human needs, the amount of hunger, suffering and conflict could be unimaginably great. Fortunately, such an apocalyptic future seems to me unlikely, most of all because there are signs that we are awakening to the challenge and still have time to intercede. It is possible to take steps to reduce the number of species lost while, at the same time, assuring that surviving species are diverse.

The case can be made that the loss of species diversity is the single greatest long-term environmental threat we could possibly face. As terrible as impacts of climate change or ecosystem collapse might be, they could potentially be overcome, or adapted to, over a few human generations. The loss of species is another matter altogether. Although the fossil record tells us that life is resilient, the restorative processes that give rise to species are painfully slow. New species will eventually evolve to replace those lost, but no time soon. People could face millions of years of hardship waiting for biodiversity to return. Better that we avoid losses in the first place.

Just as we face this crisis, we have decided to ignore the science that creates knowledge needed to inform our response. Only systematics can tell us what and how many species exist, how they are related and where they live. Only taxonomy creates a roadmap for finding our way around a complex and rapidly changing biosphere. Only systematists name species, document their attributes and create a language of biodiversity and information-rich classifications. We irresponsibly gamble with our fate as we allow systematics to be dumbed-down and decimated along with species. The reasons taxonomy—as it has traditionally been understood—is ignored have nothing to do with the quality or relevance of its science. Instead, systematics is neglected for reasons of ignorance, hubris and peer pressure.

With generally good, but nonetheless misguided, intentions, systematics has been weakened more in recent decades than at any time in its long, extraordinary history. In fact, it teeters on the brink of being redefined out of existence. Its time-honored goals are being replaced by priorities of other branches of biology. Pressing environmental issues contribute to tunnel vision, making identifying species a higher priority than getting to know them. DNA is sold as a quick fix, a way to easily distinguish among species without the trouble or expense of educating and employing professional taxonomists. Like a delirious sailor lost at sea who resorts to drinking salt water, ecologists and conservationists in urgent need of identifications eagerly accept dangerously inferior substitutes for rigorous taxonomy out of pure desperation.

In spite of what users of taxonomic knowledge may think, the primary goal of systematics is *not* making species identifiable. Although inestimably valuable, that is a by-product of basic science that seeks to discover all the kinds of life that exist, the pattern of relationships among them and the origins of their similarities and differences. Systematists formulate rigorously testable hypotheses about what species are, describe each in exquisite detail, analyze relationships and, ultimately, organize all that is learned in a *phylogenetic* classification—that is, a classification reflecting the pattern of relationships due to evolutionary history. Credible species are based on ideas and practices developed by taxonomists over centuries, including transformational theoretical advances made since the 1960s. Measuring degrees of genetic similarity cannot replace deep knowledge about species created by traditional systematics. Accepting DNA as a substitute for comparative studies of all relevant evidence, from morphology to fossils, is a fool's errand. It naïvely degrades systematics into rote procedures and stunts the growth of knowledge.

While molecular data can be parsed into characters and analyzed with the same rigor as any other evidence, arbitrary genetic distances are too often used in place of explicitly testable hypotheses.

In this book, I give a harsh assessment of the current dominance of molecular data in systematics. So let me state plainly that molecular data, as the first "new" source of evidence in modern time, is an incredibly valuable tool for systematists and that investments in molecular systematics have already reaped fantastic dividends. It is woven into the fabric of contemporary taxonomy and shall there remain.

That said, molecular data is not objectively better than other kinds of evidence. Like the others, it has strengths … and significant limitations. In groups where morphological evidence is too scarce or variable to sort out species and relationships, DNA has come to the rescue, even if questions about its results linger. It has been nothing short of revolutionary for microbiology and has challenged many long-standing ideas about relationships across botany and zoology. But with all of its accomplishments, it is no panacea. It cannot tell us much of what we need to learn.

Unfortunately, a confusion of aims and a fascination with anything new or technologically flashy have contributed to an overemphasis of molecular data at the expense of descriptions of morphology. Lip service is paid to morphology as a slow crawl toward molecular-based taxonomy continues. A recent study of wasps by Sharkey et al. is a case in point. It models a rapid, streamlined "minimal" version of taxonomy based mostly on molecular evidence. Were we interested only in telling species apart, such a diminished taxonomy might have merit. But for the lofty scientific and intellectual aims of systematics, the approach falls short. It allows us to distinguish species while knowing very little about them. The lure of easy grant money and praise from other biologists reinforces this drift toward a molecular-based approach. While DNA has worthy applications and complements other sources of data, the departure from traditional goals flirts with disaster. The bottom line is that DNA is just one of several sources of evidence for systematics; no more, and for certain important purposes, decidedly less important than others. My purpose in questioning the current preferential status of sequence data is not to reduce funding for molecular studies but to increase it for comparative studies of fossils, morphology and developmental pathways.

Handsomely funded molecular projects need no defense or promotion. So-called descriptive taxonomy, however, is in trouble. Funds for pursuing morphology-based taxonomy, including monographs and revisions, are virtually nonexistent. Such work is either carried out on the side or disguised in the trappings of some other science in order to find support. Descriptive taxonomy, done for its own sake, and to its own high standards, urgently needs to be reasserted as a high priority. We have one fleeting chance, never to be repeated, to complete an inventory of earth species and preserve evidence of evolution before millions of species have gone extinct. We must support environmental studies during this period of rapid extinction, of course. But we need not dilute taxonomy, or sacrifice the most interesting information about species, in order to do so.

Species, Science and Society is an insider's view of what is happening to the science that explores, describes and classifies species. It is a plea to seize the opportunity to discover and document the diversity of life, before it is too late; to allow molecular data to assume its proper place alongside traditional data; to reembrace the fundamentally important mission of systematics; and to dream big. By undertaking a planetary-scale inventory of species, we can possibly avoid millions of extinctions, assure that diverse species exist in perpetuity, preserve evidence of evolutionary history and discover millions of clues for creating a sustainable future.

Although the threat of mass extinction is clear and present, this book offers hope. It is not too late to complete a comprehensive exploration of species and their attributes, or to take steps to reduce the number and diversity of species that ultimately go extinct. Time remains to gather evidence of species, evolutionary history and the organization of the biosphere, and to slow the rate of species loss. Further, we have only begun to mine "ideas" found among species with which to reimagine how to meet society's needs in a sustainable manner, while protecting the natural world.

We have drawn the short straw, chosen by history as the generation that will determine the future of humanity, the biosphere and evolution. No sane person would volunteer to be in our position. Will we create baseline knowledge needed for the best conservation decisions, or continue to operate in near complete ignorance, simply hoping for an outcome as good as our intentions? Will we gather irreplaceable evidence of the diversity and history of life, or stand by as it is lost? Will we bequeath a diverse and resilient biosphere, and deep knowledge of biodiversity, to future generations, or selfishly focus on our immediate wants and needs?

Reviving the science of systematics, reembracing its centuries-old goals while modernizing its infrastructure, we can discover more species in the next few decades than in all time, before or after. We can vastly expand our understanding of biodiversity and evolution on what may be the only species-rich planet we ever have the chance to explore. And, we can make a paradigm shift, narrowing the gap between humankind and nature, ushering in an age in which species are known and appreciated; an age in which people and nature thrive in harmony.

The stakes could not be higher or returns on investment more certain. Yet, there are serious obstacles to overcome. This book draws on my career as a systematist and institutional leader to propose common sense steps to save millions of species, enrich our intellectual lives and assure continued prosperity by laying the foundation for a biomimicry revolution—a rethinking of how we live and do things, inspired by attributes of other species.

If we speak with one voice, we can demand that systematics, and a network of natural history museums, research universities and government agencies, be supported to complete one of the greatest scientific projects ever conceived: the exploration, description and classification of the millions of life forms of planet earth. No scientific enterprise promises greater or more enduring benefits to science and society, and none face such a fast-approaching deadline for action. Whether we discover the diversity, properties and history of life on earth is a now or never proposition. The future of life on earth, civilization and scientific understanding is in our hands.

Quentin Wheeler
Greenwich, New Jersey

Further Reading

Benyus, J. (1997) *Biomimicry: Innovation Inspired by Nature*, Morrow, New York, 320 pp.

Brower, A. V. Z. and Schuh, R. T. (2021) *Biological Systematics: Principles and Applications*, 3rd ed., Cornell University Press, Ithaca, 456 pp.

Cowie, R. H., Bouchet, P., and Fontaine, B. (2022) The sixth mass extinction: Fact, fiction or speculation? *Biological Reviews*, 97: 640–663.

Ereshefsky, M. (2010) Microbiology and the species problem. *Biology & Philosophy*, 25: 553–568.

Hennig, W. (1966) *Phylogenetic Systematics*, University of Illinois Press, Urbana, 263 pp.

Judd, W. S., Campbell, C. S., Kellogg, E. A. et al. (2015) *Plant Systematics: A Phylogenetic Approach*, 4th ed., Oxford University Press, Oxford, 696 pp.

Kolbert, E. (2014) *The Sixth Extinction*, Henry Holt, New York, 319 pp.

Linnaeus, C. (1758) *Systema Naturae*, 10th ed., Laurentius Salvius, Stockholm, 824 pp.

Louca, S., Shih, P. M., Pennell, M. W., et al. (2018) Bacterial diversification through geological time. *Nature Ecology & Evolution*, 2: 1458–1467.

Sharkey, M. J. Janzen, D. H., Hallwachs, W., et al. (2021) Minimalist revision and description of 403 new species in 11 subfamilies of Costa Rican braconid parasitoid wasps, including host records for 219 species. *ZooKeys*, 1013: 1–665.

Williams, D., Schmitt, M. and Wheeler, Q., eds. (2016) *The Future of Phylogenetic Systematics: The Legacy of Willi Hennig*, Cambridge University Press, Cambridge, 488 pp.

Williams, D. M. and Ebach, M. C. (2020) *Cladistics: A Guide to Biological Classification*, 3rd ed., Cambridge University Press, Cambridge, 420 pp.Wilson, E. O. (1992) *The Diversity of Life*, W.W. Norton, New York, 424 pp.

Wilson, E. O. (2016) *Half-Earth: Our Planet's Fight for Life*, Liveright, New York, 259 pp.

Acknowledgments

Ideas in this book were shaped over the course of my career, most inspired or improved by exchanges with colleagues and students. Those acknowledged should not be assumed to agree with any particular views. For comments on the book, in whole or part, I thank Caroline Chaboo, Eduardo Dominguez, Paula Ehrlich, Diana Lipscomb, Joseph V. McHugh, Kelly B. Miller, Brent D. Mishler, Peter H. Raven, Malcolm Scoble, Aaron Smith, Charles A. Triplehorn, Antonio Valdecasas, Kipling Will, David M. Williams and especially Dennis W. Stevenson. And for bringing the species-scape to life, I thank scientific illustrator Frances Fawcett.

I am grateful to the following who influenced my thinking, some in profound ways, others by sowing seeds of understanding: George Ball, Ranhy Bang, Max Barclay, George Beccaloni, Janine Benyus, Donald Borror, Thierry Bourgoin, Richard Boyd, Daniel Brooks, Paul Brown, William L. Brown, Milton Campbell, James Carpenter, Kefyn Catley, Neil Chalmers, Donald Chandler, Michael Claridge, Mary Clutter, Joel Cracraft, Peter Crane, William Crepet, Roy Crowson, Sarah Darwin, George Davis, Jerry Davis, Dwight DeLong, Eduardo Dominguez, Michael Donoghue, Henry Dybas, Malte Ebach, George Eickwort, Ginter Ekis, Niles Eldredge, Terry Erwin, Penny Firth, Jack Franclemont, Nico Franz, Vicki Funk, Ian Gauld, Henry Gholz, David Grimaldi, Andrew Hamilton, Peter Hammond, Lee Herman, Ralph Harbach, John Hermanson, E. Richard Hoebeke, Erik Holsinger, Henry Howden, Carol Hughes, David Hull, Philip Humphrey, Christopher Humphries, Robert Huxley, Daniel Janzen, Roberto Keller, William Kimbel, Sandra Knapp, Richard Korf, Frank Krell, Richard Lane, Roy Larimer, John Lawrence, Diana Lipscomb, Thomas Lovejoy, Melissa Luckow, Norman MacLeod, Christopher Marshall, Geoff Martin, Ernst Mayr, Rudolf Meier, Ellinor Michel, Doug Miller, Scott Miller, Brent Mishler, Gareth Nelson, Steve Nichols, Thomas Nicholson, Kevin Nixon, Mark Norell, Michael Novacek, John Noyes, Larry Page, James Pakaluk, Verne Pechuman, Norman Platnick, Andrew Polaszek, Richard Pyle, Peter Raven, Gaden Robinson, James Rodman, Donn Rosen, Jay Savage, Alan Savitzky, Randall Schuh, Robert Scotland, Ole Seberg, Frank Sesno, Anthony Shelley, James Slater, Aaron Smith, Gordon Stairs, Dennis and Jan

Stevenson, Maurice and Katherine Tauber, Christian Thompson, Barry Valentine, Richard Vane-Wright, Johannes Vogel, Alfried Vogler, Larry Watrous, Rupert Wenzel, Alfred Wheeler, Ward Wheeler, Donald Whitehead, Edward Wiley, John Wilkins, E. O. Wilson, James Woolley, David Young and students who suffered through my classes.

Part I

Overview

Humanity faces environmental and ethical decisions unlike any in history. They are urgent, with permanent implications for the future of biodiversity, science and civilization. Earth's species are disappearing much faster than in prehistory, erasing evidence of our past and threatening prospects for our future. We must decide what we will, or will not, do in response—before it is too late.

Unless we take decisive action, the majority of earth's species will be gone in less than 300 years. Measurable conservation goals depend on knowing what species exist and where they live, but to date we have discovered and mapped fewer than 20% of the world's animals and plants, and an even smaller percentage of its microbes.

There is, however, good news. As imagined by E. O. Wilson in *Half-Earth*, setting aside 50% of our planet's surface area, as a carefully conceived patchwork of geographic areas, could save 85% of species, possibly more. Completing a species inventory is the only assurance of having sufficient information with which to choose the right places to cede to biodiversity. Further, only an inventory can preserve knowledge of the origins of biodiversity and evidence of the primeval organization of the biosphere. As a potentially transformative fringe benefit, an inventory unlocks a treasure trove of winning ideas for creating a sustainable future.

Fortunately, systematists—scientists who specialize in the exploration, description and classification of species—have been rehearsing for a planetary-scale inventory for more than 250 years. They are primed, eager and ready. They know what must be done, and how to do it. With a taxonomic renaissance, we can resolve to explore earth's species with an urgency like that of the 1960s moonshot. The lives of millions of species depend on it, as may our own.

DOI: 10.4324/9781003389071-1

1 A Little about Molecules

Albert Einstein said that "Perfections of means and confusion of goals seem—in my opinion—to characterize our age." In respect to the mission of systematics and technological advances, he could have been talking about taxonomy. Systematists have always been early adopters of new technologies and DNA was no exception. Natural history museums and botanical gardens were quick to build molecular labs and frozen tissue banks. And systematists pioneered the use of DNA for species identification and phylogenetic analysis.

DNA barcoding continues to grow in popularity. A target region of DNA is amplified, then PCR products are sequenced. When they match a known sequence, a species' identity is confirmed. In zoology the mitochondrial cytochrome *c* oxidase subunit I gene, or *COI*, is most often used. As an important functional gene, it is present in all animals, maternally inherited as a haploid; it has a high synonymous (i.e., generally functionally silent, evolutionarily neutral) substitution rate, especially in the third codon position, enabling it to distinguish between even closely related species. In plants several plastid regions, such as rbcL, matK and the non-coding spacer trnH-psbA, serve this purpose, as does the internal transcribed spacer (ITS) region of nuclear ribosomal DNA. Barcoding assumes that less variation exists within than between species which is generally, but not always, true. As explained by Cognato et al., morphology and taxonomic expertise remain relevant and necessary if DNA-based identifications are to be reliable. Environmental DNA metabarcoding, billing itself as a fusion of field-based ecology, computation and molecular genetics, extracts DNA from samples of soil or water and then simultaneously detects the presence of many species. There remain significant issues, such as how DNA is distributed in the environment, but its potential for biodiversity monitoring is clear.

The position of molecular data in systematics is securely established. Ali et al. go so far as to credit it with saving systematics: "Now taxonomy suddenly became fashionable again due to revolutionary approaches in taxonomy called DNA barcoding," although this is not the kind of saving that systematics needs. As costs have decreased, use of DNA data has become routine. Molecular studies have impacted phylogenetics by variously confirming or upending traditional views of relationships. All of this is for the good. What concerns me is what is *not* happening. The number of species described each year is not increasing. Collections are not

DOI: 10.4324/9781003389071-2

growing to keep ahead of extinction. Taxonomists are not being educated or hired in sufficient numbers. And, phylogenetic analyses are not always translated into improved classifications.

Systematic biology stands at a crossroads. It must decide whether it will continue a tradition of scholarly excellence and synthesis of evidence or become further reliant on molecular data. The effects of neglect on taxonomy are in plain sight. Monographs and classifications are infrequently updated, collections are mothballed or orphaned and there is alarming contempt for descriptive taxonomy. Some recommend that molecular data should be mandatory in species descriptions. A few believe it should be the only evidence in order to expedite species "discovery." Describing species only at the molecular level would have serious consequences. Thousands of museum specimens representing new species, preserved in a way that makes DNA recovery problematic, would go undescribed. We would know very little about most species. And without morphology there would be no bridge to the fossil record.

Later chapters will explore broader implications for morphology that reach surprisingly deep into science, society and the future of the planet. My goal in this chapter is to stipulate that molecular data is important, welcome and here to stay. Funding for molecular studies should not be reduced, but support for other areas of systematics should be restored. The systematics community functions like a team. Bryologists take care of mosses, relying on ichthyologists to worry about fishes. Individual taxonomists should be free to choose the kind or kinds of evidence they pursue, just as they decide what taxon to study. Striving for excellence should be the one common denominator. With millions of undescribed species there is more than enough work to go around. It is counterproductive to insist that every systematist pursue any one approach, or that any one systematist pursues them all. Success should be measured in the aggregate, by the sum contributions to knowledge created, extinctions avoided and problems solved.

This chapter salutes the important and varied contributions made by molecular studies. At present, molecular work does not require propping up, but morphological, descriptive and revisionary studies need a lifeline. Fast, before expertise and millions of species have been lost. Contributions of molecular data to systematics include the following:

- *Big Data*. With increasingly sophisticated and affordable sequencing technology, molecular studies have generated an awesome volume of data. GenBank has doubled about every 18 months and, as of June 2022, contained nearly two-and-a-half billion sequences and more than 17 trillion nucleotide bases.
- *Relationships*. Molecular data has been impactful in analyzing relationships among species, providing checks and balances for morphological evidence. And, where morphology falls short, such as species complexes and unculturable microbes, it has been revolutionary.
- *Species Identification*. DNA barcodes, like UPC codes in the supermarket, are a useful way for individuals with little knowledge of species to make accurate identifications. This is possible, however, only in cases where species are known

and authenticated sequences exist for comparison. Barcodes are a powerful forensic tool, making it possible to identify mere bits of tissue.

- *Association of Sexes and Life Stages*. In some taxa, males and females look completely different, making it difficult to confidently pair up the sexes. Similarly, in organisms with complex life cycles, it is challenging to associate immature stages with adults. Molecular data eliminates the time and labor involved in rearing species to connect such disparate forms.
- *Discovering Species*. Molecular data can be used for a first-pass look at populations, flagging those suspiciously divergent from known species and thereby deserving further attention as possible new species.
- *Metagenomic Surveys*. With molecular data it is possible to use samples of water or soil to detect the presence of species without the need to collect and individually identify them. Benefits to ecology and conservation are obvious.
- *Problematic Species and Sibling Species Clusters*. When morphological characters are too variable, or too few, molecular data may sort out species.
- *Democratizing Species Identification*. In hyper-diverse taxa, like mites or parasitic wasps, an expert with access to world-class collections and libraries was previously needed to identify species. Molecular data puts the power to determine species in the hands of anyone with proper equipment and training, freeing systematists to do research rather than routine determinations.
- *Reciprocal Illumination*. Analyzing multiple sources of data is an opportunity to test each against the others. Because there is one pattern of descent, all accurately interpreted evidence must tell the same story.

As a systematist, I believe the greatest contributions of molecular data are yet to be made. It is easy to foresee a time when morphologists will use genomic data to test ideas about homology—demonstrating how divergent structures were derived from an ancestral one. It is molecular genetics that will ultimately connect the dots between genome, development and morphology, deepening our understanding of all.

As impactful as molecular data is, its newness has worn thin. It is past time that it be seen as one of several sources of data. Using multiple kinds of evidence, we become aware of mistakes; for example, when molecules and morphology indicate different relationships, at least one of them has been misinterpreted. Conflict is an invitation to revisit *both* sets of data to determine where the analysis went wrong.

DNA barcodes are suitable for identifying species for which authenticated sequences exist for comparison. Increasingly, however, they are used in efforts to discover new species. When they detect a population that differs by as much, or more, genetic distance than the average between known species, they point to what *could* be a new species and call for a closer look. However, declaring that a new species is found based on an arbitrary threshold of similarity is a self-fulfilling prophesy, not a testable hypothesis.

Subsequent chapters explain why descriptive taxonomy, based largely on morphology, is more important than ever. Why molecular data cannot tell us all that we should learn about biodiversity and evolutionary history. And why expanding museum collections and gathering diverse evidence still matters. With millions of species to be discovered and described, "minimalist" revisions, such as one by Sharkey et al., are a bad idea. They water down systematics leaving the heavy lifting of rigorous descriptions to someone else. This only increases the workload for taxonomy and delivers pitifully little knowledge at a time when we need all that we can get.

With apologies to systematists who are appropriately synthesizing molecular and morphological evidence in their work, I will take a hard line in later chapters. In my view, things have gone dangerously far favoring molecular data at the virtual exclusion of morphology. By rhetorically resisting, my hope is to make clear the importance of morphology and balance.

Beneficiaries of centuries of taxonomic progress, we take descriptive knowledge for granted and fail to see that its greatest contributions lie ahead. There could not be a worse time to turn our backs on traditional evidence or neglect the growth and development of collections. Molecular tools can, in proper proportion, elevate, expedite and advance the mission of systematics. But overemphasized, disproportionately funded, they can do enormous harm by omission. Such unintended consequences call to mind Oscar Wilde who said "Whenever a man does a thoroughly stupid thing, it is always from the noblest motives."

Most of what follows is an unapologetic defense of the traditions of systematics, a vision for completing a species inventory, a plea to pursue all evidence and a prediction of transformative contributions to society. I am frustrated that opportunities to explore and learn about species are being lost because of one despotic data source. If my anger at the neglect of descriptive taxonomy and collections boils over in hyperbole, you will know why. With renewed respect for systematics, a clear vision of the challenges ahead and the courage to confront deeply ingrained prejudices, I am optimistic that we can witness a renaissance in taxonomy and a renewed focus on collections, with unprecedented benefits for science and society.

Further Reading

Ali, M. A., Gyulai, G., Hidvegi, N., et al. (2014) The changing epitome of species identification—DNA barcoding. *Saudi Journal of Biological Sciences*, 21: 204–231.

Brower, A. V. Z. (2006) Problems with DNA barcodes for species delimitation: 'Ten species' of *Astraptes fulgerator* reassessed (Lepidoptera: Hesperiidae). *Systematics and Biodiversity*, 4: 127–132.

Cognato, A. I., Sari, G., Smith, S. M., et al. (2020) The essential role of taxonomic expertise in the creation of DNA databases for the identification and delimitation of southeast Asian ambrosia beetle species (Curculionidae: Scolytinae: Xyleborini). *Frontiers in Ecology and Evolution*, 8: 27. doi:10.3389/fevo.2020.00027

Einstein, Albert (1950) *Out of my Later Years*. Philosophical Library, New York, 282 pp.

Karim, M. and Abid, R. (2021) Efficacy and accuracy responses of DNA mini-barcodes in species identification under a supervised machine learning approach. IEEE Conference on Computational Intelligence in Bioinformatics and Computational Biology, 2021. doi:10.1109/CIBCB49929.2021.9562838

Lopez-Vaamonde, C., Kirichenko, N., Cama, A., et al. (2021) Evaluating DNA barcoding for species identification and discovery in European gracillariid moths. *Frontiers in Ecology and Evolution*, 9: 626752. doi:10.3389/fevo.2021.626752

Madden, M. J., Young, R. G., Brown, J. W., et al. (2019) Using DNA barcoding to improve invasive pest identification at U.S. ports-of-entry. *PLoS One*, 14: e0222291. doi:10.1371/journal.pone.0222291

Ruppert, K. M., Kline, R. J., and Rahman, M. S. (2019) Past, present, and future perspectives of environmental DNA (eDNA) metabarcoding: A systematic review in methods, monitoring, and applications of global eDNA. *Global Ecology and Conservation*, 17: e00547. doi:10.1016/j.gecco.2019.e00547

Sharkey, M. J., Janzen, D. H., Hallwachs, W., et al. (2021) Minimalist revision and description of 403 new species in 11 subfamilies of Costa Rican braconid parasitoid wasps, including host records for 219 species. *ZooKeys*, 1013: 1–665.

Vu, H.-T. and Le, L. (2019) Bioinformatics analysis on DNA barcode sequences for species identification: A review. *Annual Research & Review in Biology*, 34: 1–12. doi:10.9734/arrb/2019/v34i130142

Wilde, O. (2007) *The Picture of Dorian Gray*, W. W. Norton, New York, 517 pp.

Wilson, J.-J., Sing, K.-W., and Jaturas, N. (2019) DNA barcoding: Bioinformatics workflows for beginners. In *Encyclopedia of Bioinformatics and Computational Biology* (eds. S. Ranganathan et al.), Elsevier, New York, pp. 985–995.

2 Scientific Malpractice

Quietly, mostly out of sight, tens of thousands of species disappear each year as millions of others are pushed inexorably toward extinction. With fates sealed, they are what tropical ecologist Daniel Janzen calls the "living dead." Reasons for their precarious status are known to us all. Habitats are despoiled, transformed and fragmented. Non-renewable and limited natural resources are over-harvested. Industrial emissions, agricultural runoff and microplastics pollute the atmosphere, rivers and seas. Carbon dioxide increases the acidity of the oceans. Species-rich forests are clear-cut to make way for fields, plantations and urban sprawl. Invasive pests are carelessly transported threatening crops, livestock, human health and native species. As geographic and seasonal distributions of organisms shift in response to climate change, elements of ecosystems are reshuffled with unpredictable results. Faced with rapid environmental change and growing uncertainty, humankind is challenged as never before to reexamine its relationship to the natural world.

Answers to complex environmental problems begin with access to simple information. For example, what species exist and where. It doesn't get more basic than that. Yet, such information is available for only a small fraction of plant and animal species. And an even smaller fraction of prokaryotes, including single-celled bacteria, archaea and blue-green algae. So, what are we doing to redress this ignorance? Instead of commonsense investments in professional taxonomists and natural history museums, we seek technological shortcuts. We appear to be getting information about species easily and cheaply but expediency comes at a price. Identifying species divorced from the scholarship of traditional taxonomy tells us little more than a name. And without serious systematics, names and classifications have little information content.

Vernacular names may refer to one species or a number of similar ones. Dandelion and koi, for example, are common names for *Taraxacum officinale* and *Cyprinus rubrofuscus*, respectively, while "mosquito" refers to any of about 3,500 species in the family Culicidae. Often imprecise, common names are useful in everyday speech pointing as they do to characteristics shared by the organisms referenced. A "pine tree" might be any one of more than a hundred world species of *Pinus*. Still, we know that it is a tree (or, rarely, shrub), evergreen, cone-bearing,

DOI: 10.4324/9781003389071-3

Figure 2.1 World Record Insect. Dr. George Beccaloni, The Natural History Museum, London, holding the stick insect *Phobaeticus chani* Bragg. Discovered in a rainforest in Borneo, it was described as a new species in 2008 in a monograph by Hennemann and Conle. Measuring 22 inches, it is the longest insect known and emblematic of huge gaps that exist in our knowledge of earth's species. Photo: the author.

resinous, with soft wood and native to the Northern Hemisphere, with the single exception of the Sumatran pine that crosses the equator. In a world with nearly half a million kinds of plants, "pine tree," although lacking specificity, tells us a good deal. A wealth of information is tied to more precise scientific names at every level, from species to kingdom.

In a rush to embrace molecular techniques we have lost sight of the importance of describing more obvious attributes of species, making sure that species are testable hypotheses and periodically revisiting descriptions, classifications and names to assure that they keep up with current knowledge. Such core activities of systematics are foolishly seen as old-fashioned compared to an alphabet soup of DNA bases. Less celebrated, advances in theory and technology have increased the rigor and efficiency of morphology-based taxonomy, too. Far from anachronistic, morphology is as relevant to the future of science as it was to its past.

The decimation of taxonomy is taking many forms. Systematists who retire are rarely replaced in kind. Instead, researchers in supposedly sexier, more modern disciplines are hired in their place. Research grants for basic species exploration and descriptive work are almost nonexistent, unless appended to, or disguised as, a more fashionable project. Natural history museums, centers for the physical documentation of species, have lost sight of why collections exist. Attending to the preservation of specimens while neglecting taxonomy decreases the information content of collections. And, users of taxonomy, from geneticists to ecologists, have been permitted to influence priorities in order to meet their own narrow, immediate needs.

Sadly, as species go extinct, a global inventory becomes a little more manageable with each passing year. For the first time, species may be disappearing faster than they are being discovered. Even existing knowledge of species, accumulated over centuries, is in trouble. In the absence of periodic review and testing, it is slowly decaying, becoming less reliable. And the most fascinating, complex and surprising outcomes of evolution, found by comparative studies of morphology, fossils and developing embryos, are slighted as we pursue an expedient, but less informative, molecular alternative.

As the popularity of molecular evidence has increased, the decline of other aspects of systematics has hastened. Experts devoted to the description and classification of species are written off as relics of a bygone era, a pitiful time when molecular data didn't exist and scientists were driven to actually examine specimens to find evidence. OMG! Questions about *how* nature functions, from genes to ecosystems, have a monopoly on status and funding in biology. Success comes easily to those willing to go along with the crowd, to betray the long-established mission and goals of taxonomy in order to conform to politically correct experimental and molecular approaches. But for systematics, this is anathema. It was always possible to keep our eyes on the prize and simply add molecular data to the existing array of evidence, as a good number of taxonomists thankfully have done. At scale, this sensible synthesis would require that funds be shared equitably among various kinds of taxonomic research. Unfortunately, respect for alternative approaches is conspicuously rare in molecular circles.

When I became director of the division of environmental biology at the National Science Foundation, I was informed by one of my senior program officers that we received so few monograph proposals because there was no longer interest in descriptive taxonomy. The field had moved on, he told me, and the community now wanted support for molecular studies instead. Coming from this community, I knew this was not the case. I knew, also, that my colleagues are smart and had simply ceased writing proposals for revisions and monographs that they knew would never be funded. To resolve our dispute, I fenced off 3 million dollars for a special competition. A requirement to compete for these funds was the inclusion of a taxonomic revision or monograph, the epitome of descriptive taxonomy, in the plan of work. We were flooded with proposals. Money, and the desire to be accepted by experimentalists, has elevated the idea of a molecular-based taxonomy more than a rational assessment of what we ought to know and how best to learn it.

It is a disservice to science to perpetuate an artificial distinction between systematics and taxonomy. The former is often defined in a way that includes evolutionary aspects of the field, such as studies of species status and relationships. The latter is frequently restricted to rules and practices of naming and classifying, falsely perceived to be boring, arbitrary and pedantic. Systematic biology consists of both evolutionary considerations and rules, each informing the other. Thus, I join Schuh and Brower in rejecting this division that did not exist for most of the history of systematics:

> Systematics is the science of biological classification. It embodies the study of organic diversity and provides the comparative framework to study the historical aspects of evolution. Systematics—what is often called taxonomy—as currently practiced has its beginning in the work of the Swedish botanist and naturalist Carolus Linnaeus (Carl von Linné), and his contemporaries, in the mid-eighteenth century.

Because systematists analyze relationships in order to make classifications *phylogenetic*, and use species and taxon names as shorthand for scientific hypotheses, this supposed distinction is a false one. Having tired of seeing the word taxonomy used as an excuse to deny support to fundamentally important aspects of the science, I have resolved to no longer participate in this sophistry. I use the terms taxonomy and systematics interchangeably for one and the same science in all of its phases and activities. The availability of molecular data has been used to magnify this misguided distinction, further tarnishing the prestige of descriptive taxonomy. We could not imagine a competent geology without detailed descriptions of the properties of rocks and minerals, or of their distribution in strata. Yet, we are increasingly content with the idea of using molecules to tell species apart, ignoring most of what makes them distinctive and worth knowing. Systematics suffers when any of its parts are ignored. In an age of mass extinction, we need systematics in all its phases, with support for descriptions, phylogenetic analyses and classifications. This semantic trickery has done great harm to systematics as a cohesive science. No one part of systematics, or source of evidence, is as useful when others are ignored.

DNA, like any evidence, can be used appropriately or misused. When overall similarity is the basis of inferred relationships, as opposed to teasing evolutionary novelties out of sequence data, the conclusions are phenetic, not phylogenetic. Phenetics was discredited in the 1970s when it was realized that degree of similarity does not correlate with closeness of relationship. A crocodile and lizard look more alike in most respects than either does to a bird. Yet the croc and bird have shared-derived characters not present in lizards, indicating that they share a more recent common ancestor.

Those intent on replacing intellectually rewarding taxonomy with a formulaic, molecular-based alternative, bereft of aims and insights worthy of the rich history of systematics, upset me greatly. To paraphrase Howard Beale in the film *Network*, it makes me mad as hell, and I'm not going to take it anymore. After nearly 50 years as a systematist, I am furious that the science I love is being marginalized, its grand mission forgotten and great achievements disrespected. I am disappointed that the focus has turned away from understanding species and their amazing attributes to merely sequencing their DNA. I am disenchanted with natural history museums that have lost sight of why the collections in their care were assembled, and what they are good for. I am disheartened that science priorities are dictated more by money and fads than pressing needs and obvious opportunities. I am astonished how often we permit technology to become an end unto itself, rather than a means to knowledge. And, I am incredulous that traditional aims of taxonomy, more relevant today than ever, are ignored for the first time in centuries.

Because systematics is so important to science, society and our humanity, it would be a mistake to neglect its goals and needs at any time. But to do so now, in the middle of a biodiversity crisis, as millions of species face near-certain extinction, is nothing short of scientific malpractice. Our inaction to explore, describe, name and classify species flirts with damage to our economic and ecological security, and eliminates future options for adapting to changing environmental conditions. And, no less importantly, it diminishes our humanity by limiting our understanding of the diversity and origins of a biota and evolutionary history of which humans are part. As we confront the greatest extinction event in 65 million years, we are unilaterally disarming by neglecting systematics.

Often with good intentions, scientists and administrators who influence such matters ignore systematics. Denied recognition, support and positions, their theories purged from curricula, systematists are dismissed for the sin of refusing to conform to what has become a standard template for experimental biology. I am writing this book as a plea to return to our senses and support a taxonomic renaissance, before it is too late. Before so many species are gone that our understanding of evolutionary history is permanently fragmentary, our grasp of the biosphere hopelessly spotty and our options for the future few. Exploring species gives us the best chance for measurable conservation success. It preserves evidence of species diversity and evolution to enrich our intellectual lives, deepening our understanding of ourselves and the world we live in. It assures continued access to natural resources, agricultural productivity and ecological services. And, it points us in the direction of a sustainable future for civilization and biodiversity.

My goal is to convince you that all species—and the science of systematics—deserve your respect and support. That the hubris and greed that deny funds to descriptive taxonomy are horribly misguided. Only the restoration of systematics as an independent science, integrating all relevant evidence about species, will assure that our response to the biodiversity crisis is as effective as it can be, that we understand the origins of humankind and biodiversity and that we retain as many options as possible for adapting to present and future challenges of survival on a rapidly changing planet.

If we are serious about conserving biodiversity, enjoying natural resource security, innovating our way to a sustainable future and understanding how this miraculous planet of ours, with its staggeringly diverse life forms, came to be, then we must complete the elementary tasks of discovering, describing, naming and classifying species by supporting the science of systematics.

Further Reading

Hennemann, F. H. and Conle, O. V. (2008) Revision of Oriental Phasmatodea: The tribe Pharnaciini Gunther, 1953, including the description of the world's longest insect, and a survey of the family Phasmatidae Gray, 1835 with keys to the subfamilies and tribes (Phasmatodea: "Anareolatae": Phasmatidae). *Zootaxa*, 1906: 1–316.

Janzen, D. H. (2013) Latent extinction—The living dead. In *Encyclopedia of Biodiversity* (ed. S. A. Levin), 2nd ed., vol. 4, Elsevier, New York, pp. 590–598.

Schuh, R. T. and Brower, A. V. Z. (2009) *Biological Systematics: Principles and Applications*. 2nd ed. Cornell University Press, Ithaca, 311 pp.

3 The Science of Species

What are species? Plants lovingly tended in a garden. Rare birds proudly noted in a life list. The profusion of life in a tropical rainforest. Inspiration for painters, poets and writers of song. Crickets serenading from the hearth. Mystique of wilderness from which we emerged, and back to which we are drawn. Dreaded diseases, parasites and pests—and source of means to keep them at bay. Our lifeblood providing sustenance. A constant source of wonder, from colorful fishes darting in and out of corals to the blaze of sugar maples in autumn. Keys to an evolutionary history rich in details, deep in meaning. A living testament to the roots of our humanness. Survivors on a planet for which environmental change is the one constant. Models of sustainable living. Astonishing evidence of the possibilities when life arises in the vast, sterile void of space.

For most life scientists, species are "guinea pigs," subjects to be studied or experimented on with the goal of understanding some phenomenon, from intracellular mechanisms to ecosystem dynamics. For systematists, it's different. Understanding species themselves, their attributes, relationships and distributions, *is* the goal. Clades are branches in the tree of life. When clades are given names, such as a genus or family, they are higher taxa. When they include an ancestor species and all (and only) its descendant species, they are *monophyletic*. The intention in modern taxonomy is that all taxa be monophyletic. Results of systematics research include species descriptions, cladograms (branching diagrams depicting relationships, sometimes called phylogenetic trees), scientific names and classifications. Systematists employ sophisticated theories and methods, developed for their particular kind of research, that are observational and comparative. Without time travel, directly observing, much less experimenting with, events in the past isn't possible. While the purview of most biologists runs deep in pursuit of learning how life functions, that of taxonomists is fantastically broad, unconstrained by the shackles of controlled experiments or the confines of a laboratory, field site or lifetime.

At the same time that scientists quibble over precise definitions, each of us develops an intuitive understanding of species from an early age. Experience tells us that species are distinct *kinds* of living things that reproduce to give rise to more individuals of the same kind. It seems not to trouble a child that breeds of

DOI: 10.4324/9781003389071-4

dog, from Chihuahuas to Great Danes, look incredibly different in size, shape and color, yet are instantly understood to be one and the same kind of animal in spite of their differences. They are content to tell them from cats by things dogs share in common with one another and not with felines, or any other species. One instinctively understands the importance of ignoring distracting variation within a species, focusing instead on one or a combination of attributes that are unique and always present. It is obvious that white and red roses differ in color, not in kind. And that, in spite of being yellow, daffodils and dandelions are different kinds of plants.

In practice, many species are so visibly different as to be immediately recognizable by anyone paying attention. The scientific justification for species, however, is a bit more involved. It demands painstakingly constructed, explicitly testable hypotheses about the distribution of characters within and among species. This, in turn, requires hypotheses about individual characters. When an arbitrary degree of DNA similarity is used to recognize species, in place of such characters, the results are not only less reliable, they are, technically speaking, unscientific. If we arbitrarily decide that any population differing by some percentage from another is a species, it is not a testable proposition. It is a self-fulfilling claim, whether they are biologically and truly distinct kinds or not. Improperly interpreted, morphological evidence can be just as superficial and misleading, of course. This is why people once thought whales were fish, before realizing that they share certain unique characters—such as mammary glands and a uterus—with other mammals. All relevant data can and should be parsed in a way that yields discrete, explicitly testable characters.

In the known Universe, earth stands alone with a species-rich biosphere. One among billions and billions of planets, it is statistically inconceivable to me that life does not exist on other worlds. Perhaps many of them. This, however, is no reason to take life on earth for granted. It is entirely possible that humans never set foot on another inhabited planet or that, if we do, we find no more than a few single-celled microbes in the sediments of a tepid, shallow sea. While it is likely that other species-rich planets exist, including those with complex and intelligent life forms, interstellar distances make the chances of reaching one remote at best. You could say that the odds against it are astronomical. With thousands of species going extinct each year, the time remaining to explore life on our own planet is growing short.

Imagine that, at some future date, we do reach another planet with diverse life forms only to discover that its species are rapidly going extinct. What would we do? I suspect that we would spare no expense to organize expeditions to explore, collect and document its life forms as thoroughly as possible. That we would build museums, filled to overflowing with specimens, so that we could continue to study that world's species, biosphere and evolutionary history in spite of, and long after, their demise.

No less scientifically interesting, knowledge of the diversity of species here on earth is personal, too. The particulars of the evolution of earth species, a history different and unique from that of any other planet, tell the story of human origins, too. Attributes we think of as uniquely human are not. Seen in context, they are modifications of characters already present in earlier ancestors. From the outset, the astonishing diversity of life was painted with a surprisingly limited pallet. Every complex or unique character of a species originated as a modification of some pre-existing one; natural selection can only influence genes and characters that exist. As Norman Platnick explained, heritable attributes of an ancestor species are passed to each of its descendant species either in their original, or some subsequently modified, form. This is the essence of evolution at and above the species level, and explains the hierarchic pattern of relationship visible among species and characters, and why their history can be visualized as a branching, tree-like diagram.

The unusual way we walk, upright, didn't appear spontaneously. It is one of many variations on the form and use of limbs inherited from the first quadruped, the very same limbs transformed and used in other ways by leaping frogs, galloping thoroughbreds and soaring eagles. Surely, our intelligence indicates a unique human brain? Once again, no. Ours is an enlarged, somewhat differently wired version of the brain possessed by earlier primates whose brains were, in turn, modified from those of earlier mammals. Thus, to fully understand ourselves, to know what makes us human, we must trace our ancestry back, way back, all the way to the first single-celled "Eve" species from which we, and all other species, descended. To know the characters of other species, and the pattern of relationships connecting them, is to read the unabridged *Homo sapiens* pedigree and operators manual. To allow a species to go extinct undescribed, unrepresented in a natural history museum, is to tear a page out of that manual. Those who are distraught at the thought of being a tweaked version of an ape-like ancestor will not be happy to learn they are also a sophisticated derivative of pond scum. I, for one, think that it makes our humanity, and pond scum, all the more fascinating and precious.

Until now, we have been in no particular hurry to complete an inventory of species. The entire time that humans have been on earth, other species existed in abundance as a safety net. A constant, seemingly inexhaustible well of resources and ideas to which we could return, time and again, to solve problems as they arose. Suddenly, it is clear that the well is in danger of going dry, that which and how many species survive this and subsequent centuries are in question. It is equally obvious that we lack the most basic knowledge with which to make well-informed—no pun intended—decisions concerning conservation, natural resource management and land allocation. Given what we now know, species exploration should be among our highest priorities. A comprehensive inventory is our last chance to preserve evidence of diverse life forms and their evolutionary ties, our best hope for protecting biodiversity and an investment in problem-solving capabilities for the future.

We may no longer assume other species will be there when we need them. Unless, of course, we take actions to assure that they are. Even then, there are no guarantees. Not knowing which or how many species will ultimately disappear, the only insurance against ignorance is to discover, describe and preserve specimens of them all. Because extinction has made the number of species a moving target, in practice this means discovering as many species as possible, as fast as possible, vouchered in a global network of natural history museums and botanical garden herbaria. There are many social, cultural and economic reasons for diversity among museums around the world. Each has a unique history and local set of circumstances and challenges. Today, however, museums should be united by a common challenge: to coordinate the growth and development of collections in a way that maximizes the number of species represented and avoids excessive duplication of effort. Most of what we know about species today exists in the form of specimens collected over centuries. This positions museums to take a special leadership role in species exploration and communicating the history, status and fate of biodiversity to the public.

Demands and pressures placed on the biosphere by a growing human population are a primary driver of accelerated extinction. With so many species going extinct, much of what we ultimately know about species diversity and evolutionary history depends on specimens we collect and deposit in natural history museums for immediate and later study. Having only DNA profiles of species is not nearly good enough. They are like photographs of plants and animals taken from a distance, and a little out of focus. Unable to view them from different angles, zoom in or dissect them to reveal additional details, we quickly conclude that they tell us very little of what we wish to know. Collections, and the information they contain, will increase dramatically in value as the extinction crisis progresses.

No previous generation faced decisions like the ones we must now make. Our actions, and inactions, will influence which and how many species survive. The extent to which we understand the diversification of life. The kind of planet we pass on to future generations. And whether we preserve evidence of the elements and organization of the biosphere as a blueprint for future attempts to protect, repair or mimic it. History will judge us harshly if we do the wrong things or worse, do nothing. We can no longer deny what is taking place or afford to put our current wants and needs, much less fads, above this not-to-be-repeated opportunity to explore, and possibly save much of, earth's biological diversity.

Millions of species discoveries, cheap and easy today, will soon be impossible at any price. Irreplaceable details of our planet's biosphere are being lost faster than we are learning about them. We naïvely speak of restoring damaged ecosystems at the same time that we allow the record of their original constitution to be erased. Only we can mitigate this biological and knowledge catastrophe by exploring species and informing environmental decisions with facts. The Adirondack Mountains of New York State illustrate the point. After rebounding from an earlier era of heavy logging, they suffered a prolonged period of "acid rain" deposition that killed off fishes in lakes and took an unknown toll on soil organisms. Because an all-species inventory never existed, we can only speculate on the kinds and numbers of species that were lost. While it has been possible to "restore" the Adirondacks to a visually beautiful park, it is nonetheless a caricature of its former self and we will never know all of the plants, animals and microbes that lived there before deforestation, pollution and other assaults.

Our present, feel-good, random acts of conservation will fall far short of what we might achieve guided by an inventory. Only we can avoid gaps in knowledge by preserving specimens as evidence of former and surviving species diversity, and assure that knowledge consists of more than molecular data and photographs. Limiting knowledge of biodiversity to DNA sequences is about as rewarding as playing Scrabble with only four different letter tiles. Fortunately, I can think of several four-letter words that express my feelings quite well.

Increased knowledge of species can put measurable conservation goals within reach. An account of all species is much less a cost than an investment. It opens a portal through which we can access, enjoy, utilize and learn from living species. And, no less importantly, it creates a permanent archive of species that don't survive. Well-preserved museum specimens will be studied for thousands of years to come, expanding knowledge of species, adaptations, ecosystems and evolutionary history. The benefits of an inventory, description and classification of species are innumerable, immediate and long-enduring. The cost of failing to collect and describe species, or documenting them only with DNA, is unthinkable. Those who regard money spent on taxonomy as unnecessary at a time when conservation needs investment fail to see that celebrating species is essential to a heightened and enduring conservation *ethos*.

In light of the biodiversity crisis, and all it portends, it is shocking that we continue to ignore the one science whose mission is to explore species. We no longer prioritize educating young taxon experts to competently describe species. We are failing to expand natural history museum collections as the last refuge for biodiversity knowledge. With few exceptions, we are not pursuing the goals of systematics on institutional or international scales. And we are accepting strands of DNA in the place of a richly woven tapestry of evidence.

Unlike most scientific challenges, the exploration of species has a fast-approaching sell-by date. There will be no second bites at the apple. As techniques improve, we will continue to gather and refine molecular and morphological knowledge from museum specimens. But when extinct species are known only by DNA sequences or bits of tissue in a frozen ark, the door is forever closed to deeper knowledge.

The urgent need to respond to environmental challenges has put us in blinders. Progress in ecology, ecosystem science, natural resource management and biodiversity conservation is limited when we are unable to identify species. Thus, there is enormous pressure to prioritize simply telling species apart at the expense of taking time to get to know them. We feel good when we engage in science that is visibly directed at troublesome environmental problems, but no number of trees hugged can offset the disastrous impacts of ignoring details about species. Faced with fears of losing the ecological services necessary for our survival and comfort, and the extinction of charismatic birds and mammals, priorities can be driven more by emotion than logic. Any first responder will tell you that losing your head in a crisis and making emotional decisions flirts with disaster. It is far easier to pull at heartstrings and raise money with pictures of baby polar bears adrift on ice rafts than the harsh reality of extinction writ large. Ecology and conservation goals are top priorities, of course, but can demonstrably succeed only to the extent that basic species knowledge is pursued, too.

In a hostile takeover, systematics is being fundamentally reshaped. If not exactly in the image of other sciences, like ecology or genetics, then specifically to meet their needs. It is increasingly seen as a service, its own aims abandoned and forgotten. Ignorance of traditions and advances in taxonomic theory and practice, combined with political correctness, has resulted in the scientific community devaluing the systematics mission. If we are to meet the greatest environmental challenge in human history, then we must see that taxonomy returns to its own goals unfettered and with appropriate support. We can no longer afford to allow it to be distracted from its fundamental priorities by pop science, funding fads and the latest technology. With species going extinct left and right, it is unacceptable to settle for inferior substitutes for quality systematics, even when they appear to minimally meet needs of other disciplines at savings. This means respecting the intellectual integrity of systematics done for its own sake and to its own high standards. It is time to rediscover the primary reason to have natural history museums, which is to document species and facilitate taxonomy. And to see species exploration as the urgent "big science" challenge that it is.

We depend on taxonomic information. Species must be accurately identified for many reasons in agriculture, conservation and biological research, but determining their identities is only the beginning. Armed with a scientific name it is possible to retrieve information about the species, from a description to whatever other facts are known. But that is not the extent of taxonomy as information science. Based on relationships, phylogenetic classifications are both optimal information retrieval systems and the basis for making predictions about attributes of species not yet studied in detail. A species determined to belong to the family Rosaceae, for example, may be predicted to possess flowers with five sepals, five petals and spirally oriented stamens. Exceptions exist among the family's 5,000 species, but such educated inferences are far superior to the alternative.

When descriptive taxonomy is ignored, there is much less information attached to scientific names at every level. Because systematists have created so much information about the species that have names, it's easy to forget just how little we actually know. Those who would trade descriptive taxonomy for a molecular-based system either have no interest in details about species or assume that information will somehow be created, presumably through the good will of unsupported taxonomists.

Cyber-enabled taxonomy, or cybertaxonomy, is the fusion of traditional, descriptive taxonomy and the latest information and digital technologies. It represents the opportunity to modernize and make efficient descriptive work. Historically, the costs of publishing drawings and photographs have limited the visual communication of morphological evidence. Digital image technology has changed all of that. This is as exciting for morphology-based studies as the advent of DNA sequencing was for molecular-based projects. But digitization is a two-edged sword. Cyber tools can bring the creation and dissemination of visual morphological data into the information age or it, too, can become a distraction. When any technology is overemphasized, there is a risk of it becoming an end rather than a means. Extreme examples are suggestions that museum specimens could be discarded once libraries of digital images exist, or that development of image-recognition software will make professional taxon experts obsolete. Such images are a powerful tool, but they must be accompanied by descriptions, interpretations and hypotheses about characters.

It is unfair to blame the state of affairs in systematics entirely on other biologists who see taxonomy only as a source of identifications and names. They are, after all, simply doing their jobs. At the height of the popularity of the New Systematics in the 20th century, systematists were complicit in allowing the goals of their science to be comingled with those of population genetics, with what should have been a predictable loss of prestige, jobs and funding. Having sold out systematics once and failed to learn their lesson, taxonomists have now jumped on the latest genetics bandwagon in another bid to appear modern and cash in on easy money. It is unfeasible to complete systematics' mission to discover, describe and understand the diversity and history of life if it continues to court favor by emulating other sciences. This is not to say that DNA should not be incorporated into systematics; it already is and should continue to be. But rather to emphasize the importance of not losing sight of taxonomy's goals as such techniques are embraced. We urgently need unapologetic champions for the time-honored traditions of taxonomy. The more focused systematists are on their own mission, the greater asset they become to the rest of biology by providing more, and more reliable, information and knowledge.

This book is a call to arms. Acting immediately, we can slow the rate of extinction, preserve evidence of biodiversity and evolutionary history and save systematics from being repurposed out of existence. I am grateful and encouraged that we still have time to explore species and their history. That it is not too late to put systematics back on course to realize its potential. And that it is possible to modernize systematics while, at the same time, preserving the vision, mission and best traditions of the field. It is important to rebut the idea that taxonomy exists as an identification service. To clearly articulate its aims. And to add to and improve the tools available to systematists to do taxonomy faster and better, from morphology to molecules.

As astronomers lengthen the list of earthlike planets where a rocky surface, moderate temperatures, atmosphere and free water *might* support life, let's have the vision to explore and document the diversity and history of life on the one biologically rich planet accessible to us. This means that we stop treating taxonomy as a subsidiary of molecular genetics or service to environmental disciplines and reassert it as an independent, fundamental science aimed at knowing, understanding and classifying species. If we play our cards right, this century will be remembered as the golden age of species discovery and conservation. The time when science found the clarity and courage to confront a great extinction crisis by gathering common sense knowledge and laying the foundations for a future filled with biodiversity and possibilities. A turning point, when systematists rejected expedience and short-term profits to reembrace the tenets and aspirations of their own science. When natural history museums were networked as a distributed taxonomic research resource. The dawn of an era strengthening the intellectual, emotional and practical connections between humans and the natural world. And of a great reawakening of curiosity about species that put us on a fast track to a sustainable future.

Further Reading

de Carvalho, M. R., Bockmann, F. A., Amorim, D. S., et al. (2007) Taxonomic impediment or impediment to taxonomy? A commentary on systematics and the cybertaxonomic-automation paradigm. *Evolutionary Biology*, 34: 140–143.

Chang, K. (2001) Creating a modern ark of genetic samples. *The New York Times*, 8 May.

Danks, H. V. (1988) Systematics in support of entomology. *Annual Review of Entomology*, 33: 271–296.

Darwin, C. (1887) Autobiography. In *The Life and Letters of Charles Darwin*, vol. 1 (ed. F. Darwin), John Murray, London, pp. 25–86.

Humphries, C. J. (2001) Appendix II: Systematics and taxonomy in modern biology. In *Linnaeus: The Compleat Naturalist* (ed. W. Blunt), Princeton University Press, Princeton, 264 pp.

Nelson, G. and Platnick, N. (1981) *Systematics and Biogeography: Cladistics and Vicariance*, Columbia University Press, New York, 567 pp.

Platnick, N. I. (1979) Philosophy and the transformation of cladistics. *Systematic Zoology*, 28: 537–546.

Sagan, C. (1994) *Pale Blue Dot*, Random House, New York, 427 pp.

Scoble, M. J., Clark, B. R., Godfray, H. C. J., et al. (2007) Revisionary taxonomy in a changing e-landscape. *Tijdschrift voor Entomologie*, 150: 305–317.

Seberg, O., Humphries, C. J., Knapp, S., et al. (2002) Shortcuts in systematics? A commentary on DNA-based taxonomy. *TRENDS in Ecology and Evolution*, 18: 63–65.

Wheeler, Q. (2010) What would NASA do? Mission-critical infrastructure for species exploration. *Systematics and Biodiversity*, 8: 11–15.

Wheeler, Q. (2016) Why we need a 'moon shot' to catalogue the Earth's biodiversity. *The Conversation*, 21 April.

Wilson, E. O. (1992) *The Diversity of Life*, W. W. Norton, New York, 424 pp.

4 The Art of Survival

As my wife and I walked along a shore in the Galapagos, we came across a baby sea lion. The sun was mercilessly hot. The tiny, helpless creature stared straight ahead blankly. Unresponsive, barely clinging to life, its ribs protruded above its starving belly. Its mother had not returned from sea, leaving it there exposed, alone and defenseless. The sight made our hearts ache and we wished there was something, anything, we could do. But the Galapagos is carefully managed to remain in its natural state, without human interference. A benefit of protecting the islands' creatures, and avoiding human intervention, is that it is possible to walk among the animals and experience mammals, birds and reptiles up close. A downside is that you must leave part of your humanity on the boat when you go ashore to bear silent witness to both the majesty and maleficence of nature.

No biologist can visit the Galapagos without sensing the ever-present ghost of Charles Darwin. To walk in his footsteps, to see what he saw, is an emotionally charged and insightful experience. With barren black volcanic rock stretching as far as the eye can see, you stand on the jagged edge between survival and death. A tiny plant struggles to establish roots in a fissure in the cinder. Saline-incrusted marine iguanas bask in the sun. A blue-footed booby nests on the barren stone. The lava field, at the water's edge, is teeming with life, yet fortunes for any organism can turn in an instant, as the young sea lion was learning.

In this harsh zone of life, between sunbaked igneous deposits and the sea, it is easy to imagine Darwin's epiphany. Nowhere is the struggle for survival more clearly on display. Natural selection is a decisive, brutal, unforgiving arbiter of which adaptations give an edge, and which the opposite. Favorable mutations are rewarded with continued life and the opportunity to reproduce. Maladapted ones are weeded out quickly, removed from the gene pool.

From the dawn of life, forces of natural selection have sat in judgment of mutations. For billions of years, the efficient and effective have been perpetuated, the maladapted dispatched to oblivion. This is nature's research and development department. A place where random trial-and-error experiments are carried out, day and night, without rest or mercy. The result is a vast number of stories of evolutionary successes and failures. Not surprisingly, among them, we find millions and millions of clues with which we can develop more efficient, less wasteful and

DOI: 10.4324/9781003389071-5

more sustainable ways to meet our own needs. Why reinvent the wheel when we can simply study the results of such immeasurable experimentation and tinkering.

We can learn to live sustainably, if we choose. It was always a good idea to do so, but environmental problems now force our hand. Our lingering industrial-age approach to life is no longer tenable. We cannot continue to rely on exploitation, especially of non-renewable and limited resources. We cannot go on tolerating massive inefficiency and wastefulness on a planet with finite space and resources. We must reject the self-centered idea of living apart from nature, finding ways to live harmoniously with the natural world instead. We cannot continue to take without giving back.

Fortunately for us, knowledge of attributes of other species is pregnant with possibilities. Among earth's millions of species are a similarly great number of solutions for problems we now, or will eventually, face. Among species are role models for living with little to no waste. From the characteristics of individual species to their complex interactions in ecosystems, clues abound for designs, materials and processes with which we can make our civilization and biodiversity's existence enduring. We need only explore species, open our minds to the possibilities and translate what we discover into practical solutions.

With a revival of descriptive taxonomy, we can gain knowledge with which to avoid or mitigate many environmental crises and adapt to the rest. This will require new technologies and strategies for meeting human needs. The shortest path to a prosperous, sustainable future is biomimicry with its nature-inspired solutions. And the surest route to rapid and prolific biomimetic problem-solving is through a revival of systematics that includes a worldwide descriptive inventory of species. If we resolve to make it so, this will be a turning point, the dawning of a century in which we refine the art of survival.

Biomimicry refers to nature-inspired solutions to problems. As described by Janine Benyus,

> Biomimicry is innovation inspired by nature. In a society accustomed to dominating or "improving" nature, this respectful imitation is a radically new approach, a revolution really. Unlike the Industrial Revolution, the Biomimicry Revolution introduces an era based not on what we can extract from nature, but on what we can learn from her.

People have always practiced an informal kind of biomimicry. When Neanderthals first draped animal skins over their shoulders to insulate themselves from the cold, they were mimicking fur-bearing animals they hunted. When our ancestors saw birds on the wing inspiration struck and the seeds of aviation were sewn. Recently, biomimicry has been approached in a more deliberate and productive way. Janine Benyus has tapped nature's genius, as she calls it, seeking out exemplars to inspire more sustainable ways to live.

Benyus has blazed a trail forward, elevating biomimicry from an intuitive, *ad hoc* cottage industry to a way of thinking and a deliberate, organized pursuit of answers on a much larger scale. Her books, lectures, organizations and websites

Figure 4.1 Biomimicry on the Hoof. Adult and baby *Hippopotamus amphibius* in the National Zoo, Washington, DC, sometime between 1909 and 1931. Semi-aquatic habits and skin of the hippo served as models for buildings that are cooler in summer, warmer in winter. Also, hipposudoric and norhipposudoric acids, secreted by the hippo, are natural antiseptics and sunscreens that inspired a sunscreen product. The hippo, like every species, is a source of unexpected biomimetic inspiration, not just another pretty face. Photo: U.S. Library of Congress.

have led an awakening to the possibilities and reported impressive numbers of biomimicry success stories. Most successes come about, however, with an element of serendipity. Someone aware of a problem happens to see a potential solution in nature and connects the dots. This is only marginally more efficient than blind, trial-and-error guesswork. Such reliance on chance is a time-honored tradition, of course. Thomas Edison is said to have failed thousands of times before working out an effective light bulb. But this approach requires an indeterminate period of time before a workable solution is found. As we face pressing environmental challenges and rampant extinction, time is one thing we don't have. So, can we accelerate the discovery and development of sustainable alternatives?

We can, with descriptive taxonomy. It is time to take biomimicry to the next level, to bridge the gap between millions of biomimetic clues found among species and the designers, engineers, entrepreneurs and inventors who can be inspired to develop sustainable alternatives in rapid order. This requires three things: First, an inventory of species that places a premium on detailed descriptions of morphology and other attributes; second, organizing knowledge in a predictive classification; and third, a partnership with information scientists who can help refine databases, search engines and the translation of technical species descriptions into a language understandable to non-specialists.

When I read Ed Wilson's *Half-Earth*, I was surprised that the book ended on the hopeful idea that technology and human ingenuity may save us, just as they have so many times before. Thinking about the possibilities, from tapping clues among species to accelerating the retooling of society, I have come to share Wilson's optimism. We are recognizing the extinction crisis just in time to change our ways, possibly avoiding the worst losses and creating exciting new opportunities for humankind and biodiversity alike. Our challenge is to master the art of survival by becoming better students of nature, mining the attributes of species for ideas to inspire innovation and adaptation. The first steps are to arm ourselves with as much knowledge about species as possible and avoid losing potential biomimetic models each time a species goes extinct. Unaware where inspiration will strike, or the full range of tomorrow's challenges, it is important to document all species as fully as possible—especially those soon to disappear. An unimaginable number of biomimetic clues are a guaranteed byproduct of a worldwide species inventory that includes detailed studies of the characters of species. But the clock is ticking. Every day that we delay dozens of species go extinct, taking clues with them.

Further Reading

Benyus, J. (1997) *Biomimicry: Innovation inspired by nature*, Morrow, New York, 320 pp.

Lenau, T. A. and Lakhtakia, A. (2021) *Biologically Inspired Design: A Primer*, Morgan and Claypool, San Rafael, 95 pp.

Myers, W. (2012) *Bio Design: Nature + Science + Creativity*, Museum of Modern Art, New York, 288 pp.

Pawlyn, M. (2019) *Biomimicry in Architecture*. 2nd ed. RIBA Publishing, Newcastle upon Tyne, 176 pp.

Primrose, S. B. (2020) *Biomimetics: Nature-Inspired Design and Innovation*, Wiley-Blackwell, Hoboken, 122 pp.

5 Cosmology of the Life Sciences

Howard Evans framed systematics for the space age when he asked

> How much do we know about life on this little-known planet beneath our feet, the planet earth? We have not even approached the end of cataloguing the creatures that share the earth with us and this should be the very first step in our knowledge.

Users of taxonomy may see only its applications: identifying species, calling them by name and learning about their relationships, if only inferred from a classification. This view masks the fascinating, intellectually rewarding science behind names and classifications. The taxonomic study of a single group of species, such as a genus, takes the systematist on an incredible journey of discovery. Comparing thousands of specimens from dozens of museums, digesting every relevant publication since 1758, testing existing species hypotheses, interpreting complex characters, discovering and describing new species and analyzing relationships systematists are given an intimate view of a slice of biodiversity. Specimens they examine may come from multiple continents, ecosystems and centuries. Understanding a suite of characters or group of related species routinely requires a view spanning millions of years. Unlike most biologists who rely on one or a few species to understand a phenomenon, the systematist sees the grandeur of evolution divulged at the granularity of individual species and characters.

Among the life sciences, the perspective of systematics is uniquely broad, comparative and historical. This is why taxonomy is misunderstood by colleagues who, wedded to the experimental method and preoccupied with questions about how the world functions, fail to see the wonder in simply exploring what species exist and the story of how they, and their amazing attributes, came to be. Unlike other kinds of research, systematics is limited only by what evolution wrought. Its purview is unconstrained by the usual limitations of ecosystem, process, space or time. In fact, the scope of systematics is so broad, its implications so fundamental to our understanding of the living world, that the intellectual space that it occupies can be compared only to cosmology and astronomy, combined.

DOI: 10.4324/9781003389071-6

Figure 5.1 An Earthlike World. Artist's conception of Kepler-186f, the first confirmed earth-sized planet in the habitable zone of a distant star. The good news for exobiology is that the possibility of pools of liquid water signals a world where life might exist. The bad news is that even if we could travel at the speed of light—so far, a physical impossibility—it would take 557.7 years to reach Kepler-186f and as much time to return. For the foreseeable future, earth is the only planet where we can study species diversity and evolution. It is time to conserve as many, and as diverse, species as possible and to inventory them all. Image credit: NASA Ames/JPL-Caltech/T. Pyle.

Parallels between systematics and cosmology are uncanny. Ninety percent of all that exists in the Universe remains unknown, so-called dark matter and dark energy. Including microbes, more than 90% of species are unknown, too, a "dark biodiversity" consisting of millions of unseen species. Systematics and cosmology pursue similarly fundamental questions: What kinds of things exist? What attributes make each kind unique? And, what are the origins and history explaining the Cosmos, physical or living?

Sadly, the comparison stops when respect, priorities and investments are considered. The practical applications of taxonomic knowledge are so important that they cause users to stop short of recognizing the fantastic vision, challenges and rewards of systematics. Discoveries made by astronomers and cosmologists are celebrated in academia and the popular press, as well they should. Their science is granted wave after wave of funding for visionary projects and ever more sophisticated ground- and space-based research instruments in pursuit of the unknown. At the same time, taxonomists work in virtual obscurity, many of their research resources little changed in a hundred years. Instead of being seamlessly integrated into systematics, as it should have been, molecular data has become a distraction from taxonomy's mission in important ways. Describing

DNA sequences, in the absence of other descriptive studies, is leading us down a conceptual *cul-de-sac*. Synthesizing all relevant evidence used to be a hallmark of systematics, before the current push to rely primarily on molecular data. Minimizing descriptive taxonomy, rather than expanding it, tears at the fabric of systematics. Sorting species based on sequence data alone, we literally learn more and more about less and less.

Although similarly ancient, unlike astronomy, systematics gets a bad rap. Textbook definitions like "the scientific classification of organisms" are accurate as far as they go but make systematics sound about as exciting as a telephone directory—unless you understand that scientific names are shorthand notation for hypotheses that are, in turn, based on a fantastic body of theory and evidence. Classifications, far from being dry organized data, are a dynamic, information-rich entrée to all that is known about life at and above the species level.

Small wonder uninformed users of taxonomic information assume the discipline to be purely utilitarian. Daily dependence on taxonomy to identify species for biological research, natural resource management, agriculture, commercial trade, conservation and many other purposes obscures the exploration, adventure and discovery that creates taxonomic information in the first place. Of unquestionable importance, identifying species is perhaps the least interesting part of systematics.

A single scientific name like *Musca domestica*, the common house fly, represents an astonishing amount of research and scholarship. Since the house fly was named by Linnaeus in 1758, uncounted numbers of specimens have been observed. Linnaeus' claim that they possess a singular, specified combination of attributes, thereby representing a distinct kind of living thing, was put to the test by each of those observations—and evidence, from morphology to molecules, remains consistent with his conclusion. Untold hours collecting specimens, assessing variation within and between populations, making comparisons with related fly species, refining descriptions of morphology, filling gaps in genetic and geographic knowledge and keeping track of hundreds of scientific publications have gone into making *Musca domestica* a reliable and informative name. Multiply this research and scholarship by 2 million named species and perhaps as many as 5 million synonyms—species hypotheses that have been tested and rejected—and you begin to appreciate the achievements of systematics.

At a time when "big data" is prized, systematics is a reminder of the potency of quality data. In systematics, the potential exists to refute a long-standing hypothesis by collecting a single specimen. A seismic change in understanding may be sparked by one solitary observation. Taxonomy has demonstrated that, in the exploration of species and evolutionary history, the quality of evidence, and how it is analyzed, matters more than quantity. One complex morphological character may be sufficient, compelling evidence of relationship, such as xylem tubes in vascular plants or mammary glands in mammals. No mass of simple data, however large, is the equal of rigorous, careful comparative studies of complex characters, guided by appropriate theories, for understanding species and the astonishing contrivances created by natural selection.

Scientific names communicate impressive amounts of information, if you know how to unpack it. Even knowing what a species is *not* tells us a lot. Identifying an insect as a rove beetle, for example, would seem to tell us very little since the family of rove beetles, Staphylinidae, includes more than 60,000 species. Knowing only that it is a rove beetle, we can provisionally assume that it is a fast-running terrestrial predator with long, sickle-shaped mandibles, shortened elytra (i.e., hardened forewings), partially exposed flexible abdomen, conical-shaped coxae (i.e., the basal segment of the leg) and visual acuity. Exceptions exist among its tens of thousands of species so these inferences are not absolute. There are rove beetles, for example, that feed on fungus spores and a few, like Scydmaeninae, with elytra that cover the abdomen, but the odds are very much in your favor. Because beetles are holometabolous insects, it can be further assumed that it has a life cycle consisting of egg, some number of larval instars, pupa and adult. As an insect you know, too, that its body is organized into three functional tagmata: head, thorax and abdomen, with six ambulatory thoracic limbs. Being a pterygote it possesses two pairs of wings located on the middle and hind thoracic segments. Importantly, simply identifying it as a rove beetle means that it is not any of the other 1,940,000 named species. Not bad for one imprecise, family-level name. Identify the specimen further, say to subfamily, genus or species and doors open to even more, and more specific, information.

Understanding the laws of physics allows us to design aerodynamic automobiles, but no one supposes that physics exists, or deserves support, for this narrow application alone. We accept that, for both practical and scientific reasons, we should explore and understand how the physical laws of the Universe work. At their outset, basic science projects have no expectation or guarantee of practical applications. They are pursuits of pure curiosity that may or may not yield useful knowledge. History has taught us that this is a necessary investment in order to achieve breakthroughs.

Society supports such work trusting that, in the end, it will pay for itself in knowledge and applications that could never have been anticipated. We don't know what we don't know, so entirely new approaches to problem-solving depend on forays into the unknown, pursuing knowledge in places where we have no reason to expect to find answers. Only such shots in the dark recharge the well of possibilities from which we can withdraw entirely new solutions to existing and future problems. Species exploration is a prime example. Unanticipated discoveries have resulted in countless advances in human welfare, from crops to textiles and cures for disease.

Systematics at its best is an independent, fundamental, curiosity-driven science with its own mission, goals, theories and methods. Because its uses are pervasive in our daily lives, and bedrock for credible research in biology, it is applied taxonomy that is most visible to those outside the field. When you shop at the grocery store, lumber yard or garden center, you trust that the bottle of vanilla was extracted from seeds of orchids in the genus *Vanilla*, hard maple lumber was milled from *Acer saccharum* or *Acer nigrum*, and potted mums are indeed *Chrysanthemum* species (the hybrid mums we enjoy are primarily, but not exclusively, derived from *Chrysanthemum indicum*—a species also named by Linnaeus). While there are exceptions, the taxonomists who first discovered species were not motivated by commerce or applications, just insatiable curiosity. Unlike nuclear physics or astrophysics, whose expensive instruments and complex mathematics make them inaccessible to the average person, taxonomy suffers from familiarity. Many people and most biologists have at one time or another attempted to identify species with a field guide, perhaps forming the impression that anyone can do it. Identifying the majority of species—particularly in poorly-known and species-rich groups of insects, plants, worms, mites and fungi among others—is anything but simple, often best left to an experienced taxonomist.

When DNA can be compared to a library of authoritatively identified sequences, it is possible for someone with no knowledge of the organisms themselves to identify species. But most species are yet to be discovered, corroborated as hypotheses or sequenced for reference. As species are explored, we will learn the most if species hypotheses are based on more than molecular data.

To paraphrase Dirty Harry, you've got to know your limitations. I once had a neighbor who, knowing I was an entomologist, would try to impress me with insects he saw in his garden. He was using a popular field guide, seemingly unaware of its limitations. It was largely picture-driven, presenting images of one or a few common species in groups containing scores, sometimes hundreds, of regional species. When there are large numbers of species in a group, narrowing it to a single species by picture is risky business. I would humor him, nodding approvingly, while being far less confident about the identity of a species I had seen in my own yard.

To appreciate the fundamental science of systematics it is best to begin with the big picture. Far from a mere source of identifications, the mission of systematics is outrageously audacious:

> The mission of systematics is to discover the millions of kinds of plants, animals and microbes on, under and above the surface of an entire planet; to enumerate the combination of attributes that make each species unique; to reconstruct the pattern of relationships among species due to evolutionary history (phylogeny), integrating all relevant sources of evidence whether morphological, paleontological, embryological, molecular or other; to give each species a unique name; and to organize all that is known of species and their attributes in a phylogenetic classification serving as a general reference system for biology.

For those who think of taxonomy as biological bookkeeping, this may come as a surprise. But to those of us fortunate enough to spend a lifetime in systematics, it has always been so. Two thousand years ago, Aristotle contemplated the significance of similarities and differences among living things, wondering why some animals have blood or backbones, others not. In the 18th century, Linnaeus courageously set out to discover, diagnose, name and classify all the species in the world. Given limitations on travel, communication and access to literature and collections in his time, his discovery and naming of about 10,000 species was a remarkable triumph.

In the centuries since, professional and amateur taxonomists have continued Linnaeus' enterprise that ranks among projects we today call "big science." That description is usually reserved for things like the Large Hadron Collider or the Webb telescope: obscenely expensive instruments shared by large scientific consortia. Not generally described in the language of big science, the exploration, description and classification of millions of species, and reconstruction of their evolutionary relationships, clearly qualify, too. Just because taxonomy may advance with modest instruments, libraries and collections; its experts are scattered around the world; and its progress has been incremental over centuries, it is rarely thought of in such grandiose terms. But this is a limitation of perception, not reality. To complete a planetary-scale species inventory, it is time to reimagine systematics as big science in the modern sense, and to align its workforce, collections and instrumentation with the magnitude of the task. It makes sense to retain and build on systematics' existing strengths while modernizing its infrastructure and organizing a worldwide research effort. We must insist on a pace of species exploration commensurate with the rate of extinction. A species inventory can no longer be a slow-motion project played out over hundreds of years. Taxonomists are motivated to efficiently finish what Linnaeus started. We need only endorse the goal, provide resources and get out of the way.

"Dark biodiversity" is as much a hindrance to understanding evolution and the biosphere as dark matter and energy are for explaining the Universe. Both cosmologists and taxonomists want to discover and characterize all that exists and to probe its origins. Unlike cosmology, ignorance of species creates immediate dangers. Bad taxonomy can be a matter of life and death, as herpetologists handling misidentified venomous snakes have found out the hard way. And, millions of unrecognizable species can spell disaster for the best intentioned conservation goals. While monitoring ecosystems with a few key species is a logical practice, it is best to never forget Ed Wilson's frequent admonition that it is the little things that run the world. In a worst-case scenario, the extinction of too many, or the wrong, species could cause the collapse of ecosystems plunging humanity into a new "dark age," potentially threatening our survival. This extreme is thankfully unlikely, but the ecological principles at play are all too real. Decisions we make today will shape the course of evolution and limit the possibilities available to society for improving the human condition. In the very long term new species will arise to replace those lost, but the processes of evolution can act only on surviving species. Even in a world robbed of species richness, humans would likely continue to find ways to improve their lives. More slowly, however, than if they had access to knowledge of diverse adaptations of species.

Cosmology is important to our intellectual lives by allowing us to understand the origins of our world. Systematics plays the same role for living things. A crucial difference is that the costs of ignoring systematics can take a toll sooner. Stars and planets studied by cosmologists and astronomers will eventually disappear like species, but no time soon. With the exception of a few more pot holes, the surface of Mars will look very much the same a hundred thousand years from now. Within two or three centuries, earth's biosphere may be unrecognizable.

Cosmologists have the advantage of studying the history of a Universe that is governed by physical laws making many things as predictable as clockwork. Evolution is not so neat and tidy. The course of evolution is contingent upon random events in addition to somewhat more predictable forces, like mutation rates and modes of natural selection. On the other hand, taxonomists have the advantage of historical clues preserved in complex characters and genomes. While approaches to cosmology and systematics are necessarily different, both are driven by curiosity, historical constraints and the dream of knowing all that exists and how it came to be.

The vast number of species on earth requires taxonomists to specialize. It is simply impossible for individuals to learn more than a modest number of species in a lifetime, on the order of a few thousand at most. As a result, there is a whole lexicon of names for taxonomic specialists. You likely know specialties focused on really large groups like botany (plants) and zoology (animals). But each of these may be teased into more focused areas of expertise. Here are just a few specialists and the groups they study:

Acarologist	mites and ticks
Arachnologist	spiders, scorpions, harvestmen and relatives
Bryologist	mosses
Chiropterologist	bats
Dipterist	flies
Herpetologist	reptiles
Ichthyologist	fishes
Lepidopterist	butterflies, moths and skippers
Lichenologist	lichens
Malacologist	molluscs
Microbiologist	microbes
Mycologist	fungi
Nematologist	round worms
Odonatologist	dragonflies and damselflies
Orchidologist	orchids
Ornithologist	birds
Phycologist	algae
Pteridologist	ferns
Tardigradologist	water bears

Each specialized area of taxonomy comes with its own vocabulary for morphological structures, species and higher taxa. The taxonomic specialist is usually an accomplished field biologist, too, with knowledge of natural history, including details of evolution, habitats, geography, development, seasonality, behavior and other aspects of species.

Well-meaning efforts to address the decline in systematics in recent decades have mostly fallen short for one reason: they have focused on the needs of users to access taxonomic information, not the needs of taxonomists to create and verify knowledge. A prime example is the Encyclopedia of Life. E. O. Wilson had the inspiring vision of a web page for every species, and impressive progress has been made populating such pages. But with the focus on consumers rather than producers, it has mostly moved static, sometimes outdated, taxonomic information from printed to web pages, doing little to speed either the description of new species or critical reevaluation of existing ones. The project as a whole has failed systematics and, in so doing, its target audience, too. Had it been designed by and for taxonomists, it could have been built as a software platform with two complementary purposes. First, to facilitate the description, revision and naming of species. And second, to instantly translate conclusions into user-friendly web pages consisting of the very latest and best information. Conceived with users in mind instead, taxonomists are left with outmoded description tools and efforts are wasted moving information from journals to web pages. This is a missed opportunity for taxonomists and users alike. Whether we learn from this mistake and focus on enabling taxonomy is just one of many choices that we must now make.

Further Reading

Evans, H. E. (1968) *Life on a Little-Known Planet*, E. P. Dutton & Company, New York, 318 pp.

Parr, C., Wilson, N., Leary, P. R., et al. (2014) The encyclopedia of life v2: Providing global access to knowledge about life on earth. *Biodiversity Data Journal*, 2: e1079. doi:10.3897/BDJ.2.e1079

Tyson, N. DeG. (2021) *Cosmic Queries*. National Geographic, Washington, 312 pp.

Wheeler, Q. (2010) What would NASA do? Mission-critical infrastructure for species exploration. *Systematics and Biodiversity*, 8: 11–15.

Wilson, E. O. (2003) The encyclopedia of life. *Trends in Ecology & Evolution*, 18: 77–80.

6 Choices

As species disappear, we are confronted by unprecedented choices. Will we be hapless bystanders as biodiversity is decimated or create knowledge with which to mitigate losses? Do we leave the fate of life on earth to chance, hoping things somehow work out, or act to shape the future? Will we be silent witnesses as evidence of evolution is swept away or gather and preserve specimens as a permanent record? Do we rely on chance discoveries to create ways to adapt to a changing environment or systematically—in both senses of the word—explore species to point us in promising directions? Will we settle for minimal knowledge of species or learn as much as we can? And, do we suffer whatever consequences come with mass extinction or take steps to avoid, mitigate and adapt to change?

As science fiction writers have long speculated, we may one day, a long time from now, leave the mess we have made of our world behind to take up residence on some other planet. In the meanwhile, perhaps we could desist further despoiling this one. On the whole, our track record protecting the environment isn't very impressive. As we draw up plans to colonize Mars, we should make peace with our home planet, too, vowing to take better care of a biosphere that has survived drifting continents, super-volcano eruptions, rising seas, changing climates and meteor impacts in the past. Given the chance, it can survive us, too. Peering into the void of deep space, we should feel overwhelming gratitude to find ourselves on a planet bustling with life. We speak passionately of conservation and environmental stewardship, but our words are betrayed by our actions. How committed are we to saving species if we don't even make an effort to recognize, describe or name them?

We depend on other species in the form of ecosystems to capture solar energy, filter water, generate oxygen, cycle nutrients and sequester carbon among other things. Comprised of interacting species, each remarkably well adapted to its role, ecosystems are intricate networks. Imperfectly mimicking their functions, even with incomplete knowledge of their species, we can increase efficiencies in industry, agriculture and urban infrastructure. But system-level emulation of nature is only a tentative first step into biomimicry. Viewed at the granularity of species, millions of possibilities for sustainable living come into view, not to mention the usefulness of species themselves as untapped resources. Individually,

DOI: 10.4324/9781003389071-7

species already enrich our lives in the form of foods, medicines, timbers, fibers and more. Even knowing a small fraction of species, we have found tens of thousands of them useful. With an inventory of all species, this is just the beginning. Millions and millions of species are potentially useful for things we have not yet thought of, including teaching us to live sustainably.

I like to eat and breathe as much as the next person. All the same, I find biodiversity conservation arguments based primarily on protecting ecological services disappointingly unimaginative. Ecological services are essential for our survival, but they are not the only reason to value species enough to save them. Ecological services speak to our animal needs, appealing to the same selfish human nature that got us into this predicament. The question "What's in it for me?" is a poor foundation for an enduring relationship with nature. It implicitly devalues millions of species because they have no current practical use or visible role in the ecological services we want. When push comes to shove, and tough decisions are required, most species will not be considered essential enough to save under the criterion of ecological services.

The Pollyanna explanation that every species occupies a unique niche in an ecosystem and, therefore, should be protected in order to safeguard services generated by the system as a whole is a noble sentiment that simply does not hold water. The reality is that systems with far fewer species are, at least potentially, capable of continuing the services we care about. It is possible to poke a bunch of holes in the web of life while leaving enough strands intact for the narrow purpose of fulfilling human needs.

This cynical, human-centric, ecological argument puts more than "non-essential" species at risk, it threatens to impoverish our humanity, too. We are capable of so much more than focusing on our physiological needs, of feeding our mind, spirit and imagination as well as our body. This, however, requires that we see species as more than commodities, bags of DNA or interchangeable gears in an ecological machine. Intellectual curiosity, ethics and aesthetics are just a few of the lenses through which species deserve to be seen.

Naming species is a crucial step toward honoring them. It acknowledges their existence and inherent worth apart from what they might do for us. Species names, like people's names, individuate. My surname tells you I am a member of the Wheeler family, but which one? My given name, Quentin, answers the question. Two-part species names serve the same function. The name *Quercus rubra* tells us that a tree belongs to the genus of oaks, *Quercus*, and that this particular one is the red oak. This name has meaning because it is tied to hypotheses about the identity and characters of red oaks and oaks generally. Left unnamed, reduced to a mere statistic or DNA sequence, species become faceless, abstract concepts. This allows us to remain detached from the ugly tragedy of their extinction. Reduced to numbers in a table, or sequences in GenBank, does it really matter whether there are 5000 species of mammals or 4000? But look an elephant or sloth in the eye and it is far more difficult to be unmoved by the prospect of its extinction. Each species is a distinct kind, a remarkable, unique creation of evolution, a winner in the lottery of life deserving recognition and respect. Each species has its own attributes, history and things to teach us. Coming to know a species, calling it by name, makes indifference to its fate far more difficult.

E. O. Wilson once compared species to great works of art, each a masterpiece in its own right. Extending his analogy, we would be saddened to hear that several pieces in the Louvre had been destroyed. But the news would hit much harder if they were the Mona Lisa, the Wedding Feast at Cana and Venus de Milo, rather than works we had never heard of. In almost every context—phylogenetic relationships, role in ecosystem, attributes, association with humans—*which* species go extinct matters. The ideal, of course, would be to save every species and to the extent possible this ought to be our goal. But the current situation makes that ideal unattainable. Environmental changes and human actions are driving too many species toward extinction too rapidly. The next best thing to saving every species is to put our thumb on the scales of selection tilting the odds in favor of keeping a more, rather than less, diverse biosphere.

We feel special that we are the only living species of hominid, *Homo sapiens*. But this was not always so. Anthropologists continue to fill gaps in our family tree, discovering extinct relatives who have not been gone all that long in geologic time: *Homo erectus*, *Australopithecus africanus*, *Homo floresiensis*, *Homo habilis*, *Paranthropus robustus* and *Homo neanderthalensis* among them. Rather than gloating that we are so special that there is only one of us, perhaps we should worry that hominids have not been more successful measured in numbers of species. If we are so special, why does Aves, consisting of feathered and literally bird-brained descendants of dinosaurs, have thousands of species and hominids only one? And how is it that insects with ganglia—nodules along a string of nerve tissue, stretching from head to tail—rather than a centralized brain, have millions of species and hominids just one? Now, who's special?

Humans are, of course, remarkable in other ways good and bad. Uniquely capable of wreaking havoc on the biosphere on one hand, nonpareil in intellect, curiosity, creativity and the ability to positively shape the future on the other. We owe it to ourselves to pursue an understanding of biodiversity and its origins that is deeper than one data source and more meaningful than meeting practical needs.

Ecosystem science, without intending to do so, has led the way robbing species of individual worth. Measuring inputs and outputs of systems, individual species become invisible cogs in the machine. For the function of the system, what matters is that a prescribed weight of leaves is consumed each day. Whether the herbivore is a caterpillar or beetle really doesn't matter. Given equal appetites, they are interchangeable. At a seminar in the early 1980s, I recall a theoretical ecosystem scientist proudly proclaiming that even if every species at his study site could be identified, he wouldn't bother. His mathematical model was so good at explaining how the system worked that knowledge of species was an unnecessary detail.

This obsession with function, at the exclusion of curiosity about species, contributes to an increasingly narrow, simplified and dangerously anthropocentric view of biodiversity. What happens when this selfish perspective is taken to its logical extreme? We don't need to speculate. Some have gone there already. An essay by Alexander Pyron in the *Washington Post* unmasks the depth of this narcissistic world view:

> But the impulse to conserve for conservation's sake has taken on an unthinking, unsupported, unnecessary urgency. Extinction is the engine of evolution, the mechanism by which natural selection prunes the poorly adapted and allows the hardiest to flourish. Species constantly go extinct, and every species that is alive today will one day follow suit. There is no such thing as an "endangered species," except for all species. *The only reason we should conserve biodiversity is for ourselves, to create a stable future for human beings*. Yes, we have altered the environment and, in doing so, hurt other species. This seems artificial because we, unlike other life forms, use sentience and agriculture and industry. But we are a part of the biosphere just like every other creature, and our actions are just as volitional, their consequences just as natural. Conserving a species we have helped kill off, but on which we are not directly dependent, serves to discharge our own guilt, but little else.
>
> (italics mine)

The elimination of polymorphism by extinction is the engine of evolution, along with the removal of entire species. For example, the order Equisetales has only one surviving genus, *Equisetum* (horsetails), but in the geologic past had about 60 genera, including some species that grew over 30 meters in height. The fact that extinction is a natural phenomenon in no way justifies indifference to what humans are doing to biodiversity.

We cannot foresee the needs or wants of civilizations who follow us. Species of direct impact on our lives today may not be the same as those valued centuries from now. It is dangerously arrogant to think, with knowledge of fewer than 20% of species, and no idea of the challenges that will arise in the future, that we know enough today to decide which species deserve to live by Pyron's selfish criterion of direct benefit. Before someone invented Roquefort cheese, no one would have imagined a lowly mold, *Penicillium roqueforti*, to be either useful or worthy of conservation. Armed with baguette and Bordeaux I for one consider it of direct benefit to humankind, as evidenced by my happiness and that of *les Français*.

Whether Pyron admits it or not, the human impact on biodiversity is unnatural. It differs in kind from impacts of every other species, with the possible exception of Neanderthals or other pre-modern hominids. Human actions, some deliberate, others made in ignorance or disregard, lie outside the normal governance of species diversity and are, at best, akin to the kind of artificial selection that has produced cattle breeds that never would have existed without our intervention. This is not comparable to the actions of other species that result in extinction. Non-humans impact a rather small circle of species with which they interact, such as prey, predators and parasites. Humans alone affect so many species on a global scale. This is a deliberate and avoidable impact, decidedly unlike that of any other species. Further, we can anticipate consequences of our behavior. As sentient beings, we should feel justified remorse at what we have done when a species disappears as a consequence of our actions. With our awesome powers of reason, creativity and technology come ethical responsibilities toward species unequipped to deal with our actions. In most corners of the globe *Homo sapiens* is an invasive species. As house guests we should be better mannered.

Looking at the big picture, it could be argued that we are or will be, sooner or later, dependent, in ways large or small, on many other species—so many that it is foolish to discount any at this point in time. It is impossible to anticipate which species will demonstrate their worth as the environment changes, and the height of folly to think we know which species we might value in the future.

I believe that humankind is better than its occasional fits of blind greed and stupidity. That we are capable of seeing worth in species irrespective of what we might get from them. Spending countless hours as a child collecting, culturing and observing protists, I came to deeply admire them. This experience eventually led me into a career as a taxonomist and it never once occurred to me to ask whether such humble species deserve to continue living. Boundlessly fascinating, with an evolutionary history much longer than our own, of course they do.

There is evidence of our better selves all around. Philosophers ponder the ethics of our domination over other species. Theologians weigh our responsibility to the Maker to care for His Creation. Artists and poets create great paintings and phrases, inspired by the beauty of other species. Tourists flock to national parks to experience species diversity on a landscape scale. Birders rise before dawn for the mere sight of a rare species. Like an ultimate dive into genealogy, we have an innate yearning to learn how our own species is related to others.

Evidence of genuine concern for other species is unfortunately limited. It is not enough to focus on cute and cuddly animals and showy flowers, we must sense a connection to all species. Anyone with their humanity intact feels for an abandoned, starving baby seal. It is worth asking why we are so callous toward a flat worm or carrion beetle in similarly dire circumstances. Each is a wonder of nature, a remarkable survivor in the struggle for life, with something fascinating and unique to tell us if we are wise enough to listen. Each is related to us, more or less closely. Who are we to reduce the amazing diversity of life to a beauty pageant? Or draw an arbitrary line between the worthy and the expendable? In the past, that line was drawn between humans and non-humans with indifferent cruelty shown to other animals. Mass extinction dictates that the line must be redrawn, but where, and by what criteria?

Perhaps naïve, I have higher expectations for humans than seeing biodiversity as a storehouse of goods and ecosystems as a source of services. This is the same self-serving arrogance that led us to lord over nature, a dominance that we abused horribly. By naming and learning about species, we recognize, individuate and honor them. Indulging our curiosity, we pull the curtain back on an evolutionary history more surprising than fiction. Through discovering the grandeur of biodiversity, we gain humility and cease to measure species only by their usefulness.

The ancient idea of a *scala naturae* lingers, the notion that life has evolved in a linear, upward trajectory toward ever more advanced or perfected life forms with humans, of course, at the pinnacle. In reality, humans are just one of millions of terminal branches on the tree of life. We are "advanced" in some ways, not so much in others. When it comes to photosynthesis, for example, we are not nearly so highly evolved as a tomato plant. And, in the web-spinning department, arachnids put us to shame. It is fitting that we come to know phylogenetic history so that we see ourselves in context, with less vanity and sobering realism.

We should seek more from life than creature comforts. Holding ourselves to a higher standard, we can aspire to live, not just survive. It is by examining our ethics, morals and relationships to other species, indulging our curiosity, admiring beauty in nature and retaining a childlike sense of wonder that we find deeper meaning in life. We need not choose between human-made environments in which we live, work and play, from cities to cyberspace, and being connected to nature. With occasional visits to wilderness and museums, and by engaging in contemplative thought, we can have both.

The extinction crisis is a test of which species survive. It is also a test of our humanity, a measure of what we're made of. It will show whether we are as greedy and self-centered as our worst past actions suggest, or if we have the moral clarity and intellectual capacity to respect other species for their own sake and accept our rightful place among them—and they around us. We have the option to care for other species and, in the process, create a more secure and fulfilling future for ourselves. What we do in response to rampant extinction is an opportunity to look into the mirror and see who we truly are. Robert G. Ingersoll said, in a quotation often attributed to Abraham Lincoln, "Nearly all men can stand adversity, but if you want to test a man's character, give him power." As we decide which and how many species live or die, and whether we seize the opportunity to explore a biosphere about to be decimated, the unprecedented power we are granted will reveal our collective character.

Humans are a walking contradiction. The same *Homo sapiens* that creates great sculptures and literature defiles subway cars with graffiti and fouls the air and water. The same human race that established the national park system clear-cuts forests and harvests sea fishes to the brink of extinction. The same capitalism blamed for industrial waste lifts people from poverty and creates the wealth necessary for the luxury of doing environmental good. We need not sacrifice quality of life or personal aspirations to accept the opportunity to treasure, protect and enjoy the diversity of life.

The more we understand biodiversity, the greater the opportunities to call forth the better angels of our nature. Acknowledging that we have arrived on stage, for what ought to be a minor supporting role, deep into the plot of an extremely long evolutionary play, should give us the humility with which to reassess our relationship to the rest of the cast. Seeing the big picture, we should be motivated to examine our place on this planet, honestly assessing our past actions and boldly challenging ourselves to correct course.

Just as altruism toward the less fortunate makes us better people, compassion and respect for all species elevates our humanity. While we should make cities aesthetically pleasing places to live, we only need get out of the way to allow nature to remain astonishingly beautiful. We have the choice to do no further harm to the Eden we inherited. Beyond conserving and celebrating what remains of the natural world, we owe it to posterity to conserve, too, evidence of what is being lost on our watch. This requires us to rise above immediate interests to complete a massive task of enormous, enduring worth. By embracing systematics, running toward the flames of extinction, knowledge will be created to make the world better. Many things hang in the balance, but three seem particularly important to me.

History

We cannot fully understand ourselves, or the world around us, without learning about the sequence of events responsible for the diversity of species. Darwin's theory of evolution changed how we see ourselves. But recognizing evolution was the first step on an adventure of discovery, not the end. We have only begun to explore the results of evolution, much less trace the steps of life's diversification. Fossils, molecules, developmental pathways and comparative morphology each tell us essential parts of the story.

Phylogeny is like a great jigsaw puzzle. Put millions of pieces together and see the big picture of evolution. So far, we have laid hands on fewer than two out of ten pieces. Worse, pieces are being destroyed faster than we are gathering them. Unless we collect specimens and observations of the rest, we risk permanent gaps in the picture. Because few species leave a fossil record behind, millions of pieces will never be added to the puzzle unless we collect them now.

To push the analogy further, like jigsaw pieces, each species occupies one particular position. The characters of a species, like the shape and visual clues of a puzzle piece, allow us to see where it fits. Even if DNA can put all species in their place, doing so in the absence of morphological knowledge would be like solving the puzzle face down. We would never see the picture that we had set out to reconstruct. It would be like trying to appreciate why *North by Northwest* was such a good film by studying Hitchcock's cameras and sound gear. They show us the mechanics behind the movie but completely miss the reasons that the film is worth our attention.

Suppose we eventually reach another planet with many and diverse species, a remote possibility at present. Its life forms and evolutionary history would not be those of our planet. We owe it to ourselves to understand the diversity and history of life here, on earth, both for what it can do for science and human welfare, and as a benchmark. After discovering biodiversity on some other world, our very next impulse would be to ask how its diversity, history and evolution compare to those of earth. It would be more than a little embarrassing to admit that we had missed the opportunity to explore life on our own planet because we were too busy dealing with immediate problems on one hand, and reaching for the stars with the other.

Conservation

Unaware of what or how many species exist, where they live or what they do, how can we possibly make the best conservation decisions? Wilson's *Half-Earth* proposal is an ideal combination of science and common sense, impressively ambitious, yet so simple in concept that it could actually work. Wilson estimated that, by setting 50% of the surface area of the planet aside for biodiversity, we could save about 85% of all species. Assuming that we can overcome enormous hurdles related to politics and economics, the number and diversity of species conserved by the Half-Earth Project will still depend on selecting the right combination of geographic places. This, in turn, requires knowledge of what species exist and where. Until we complete the elementary task of mapping the species of the biosphere, conservation will remain more slogan than plan, a collection of feel-good acts that leave the fate of biodiversity to chance. Fortunately, there are signs of hope, such as the Half-Earth Project's ambition to map species at a resolution of 1 km as a start.

Quality of Life

Carl Sagan put the importance of valuing earth into perspective,

> The Earth is the only world known so far to harbor life. There is nowhere else, at least in the near future, to which our species could migrate. Visit, yes. Settle, not yet. Like it or not, for the moment Earth is where we make our stand.

The world is changing, fast. Unless we act to mitigate, stop or reverse current environmental trends, earth will be a very different place a century or two from now. Biodiversity decimated. Ecological webs tattered and frayed. Climates changed. Natural resources exhausted. Landfills overflowing. Oceans, waterways and the atmosphere fouled. Wilderness plowed under. Our best hope is to avoid or minimize environmental problems while creating new ways to meet our needs, with fewer and less damaging impacts on the natural world. Even though currently pressing environmental issues were created, or at least magnified, by humans, earth is a dynamic planet and dramatic change is an ever-present possibility, a question of when, not if. This is not to suggest that a background of "natural" climate and environmental change releases us from the responsibility of facing up to our own impacts, but to say that whether problems arising are our fault or not, we should be prepared for every eventuality.

We can decide whether we face an uncertain future with ignorance or knowledge. Investments in knowledge today can avoid some catastrophes and guide us through others. The acceleration of extinction has been accompanied by a reduction in our capacity to explore species, including an erosion of expertise, programs and sources of support. Scientists and institutional leaders are smart people. So, you may be thinking that there must be a good reason that systematics has been marginalized. Perhaps it suffers from some debilitating shortcoming as a science. After all, there are those who dismiss it as stamp collecting. But scientific weakness can't be the reason. Beginning in the 1960s, systematics underwent a theoretical revolution making it as rigorous as any science. Maybe advances in systematics are limited by technology. Again, not so. There are off-the-shelf technologies that, with a little tweaking, could immediately increase the pace of species description by an order of magnitude.

The reasons that systematic biology is neglected have to do instead with misunderstandings, political correctness, greed and hubris. Hardly motives we like to attach to scientists or that we should be proud of. The good news is that these can be overcome. Systematics is so fundamental to understanding the world, so impactful on our daily lives and so important to the creation of a sustainable future that it will eventually resume center stage whether we act today or not. But such shifts of scientific priorities can take a generation or more to effect and by then we will have lost an enormous number of species.

First and foremost, we need the systematics community to resist the forces of conformity that have it in retreat and reassert the inspiring vision of its mission. Finding the courage to lead, a few research universities and natural history museums can become institutional catalysts for change. This is also a situation that could be influenced by a grass roots movement. In the U.S., for example, were the public to demand of Congress that a species inventory be a priority, things could change overnight. Fully apprised of the facts, I believe that the public would get behind an effort to learn all we can about disappearing species, give conservation its best chance for success and tap biomimicry for new options. To any thinking person, it is obvious that we cannot remain dependent on unsustainable means to meet our needs. This book makes the case for a taxonomic renaissance and worldwide species inventory. It is a plea to rediscover a sense of wonder about biodiversity and evolution and strengthen our connections to the natural world through knowledge.

It would be easy to be discouraged by the challenges we face. Instead, we should be grateful and inspired to action. We are privileged to have the chance to explore the most biologically rich planet in the known Universe. To discover more kinds of plants and animals in the next 50 years than in all time, before or after. We can be the ones with the foresight to prepare humankind to meet unprecedented environmental challenges and the wisdom to preserve evidence of our biological heritage. We do an enormous disservice to ourselves if we value earth only for what we can extract from it. That is, after all, how we painted ourselves into this corner. In doing so, we ignore the things that most impressively separate us from the beasts: an irrepressible curiosity about the world, a longing to explore new frontiers and a drive to make things better.

As we explore species, we will stand in awe before their diversity and ask ultimate questions about life. How have improbably complex, wondrously detailed morphological structures come about from simpler body parts? How was the biosphere organized before humans began disrupting things? Why are we, and every other species, as we are? What sequence of transformations explains the astonishing diversity of life? The answers to these, and many other questions, are found in the comparative study of species. Knowledge of the history of life, summarized in phylogenetic classifications, enriches our lives and deepens our understanding of the planet.

Dire predictions about the loss of biodiversity and ruination of the biosphere are demoralizing. There is no denying our tragic circumstances, yet it is vitally important that we not give up. Instead, we must salvage what we can in respect to biodiversity and knowledge. In this book I offer the prospect of a better future by doing the right things. It is not too late to revive systematics in its full richness, put the brakes on extinction and reset our relationship to the natural world. I hope that we will be remembered not as a generation that stood by as the planet was changed, but one with the courage and vision to explore, learn and act. Let us dare to honor other species with names and familiarity, question science priorities in the face of mass extinction and choose a future rich in species, knowledge and possibilities.

It is time to reimagine systematics as a powerful fusion of traditional goals and modern capabilities. Reembracing its timeless mission, we can restore taxonomy's prestige, modernize its infrastructure, streamline its workflow and inspire a new generation of species explorers. Exobiology can wait, if need be. To our knowledge only earthbound species are in imminent threat of extinction. We are about to be seriously tested. Do we have the ethical fortitude to respect other species enough to discover, describe and name them? Do we have the discipline to cease despoiling the planet by reimagining how we ourselves live? Do we have the wisdom to preserve evidence of the diversity and history of life on this planet before it is too late? And, do we possess the largesse to partition the world in such a way as to permit biodiversity to survive? Let's hope we pass these tests.

Further Reading

Alcamo, J., Ash, N. J., Butler, C. D., et al. (2003) *Millennium Ecosystem Assessment: Ecosystems and Human Well-being*, Island Press, Washington, 245 pp.

Asimov, I. (1979) *A Choice of Catastrophes: The Disasters that Threaten Our World*, Simon and Schuster, New York, 377 pp.

Half-Earth Project Map (2022) http://half-earthproject.org/explore-maps/ (Accessed 9 January 2023).

Sagan, C. (1994) *Pale Blue Dot*, Random House, New York, 427 pp.

Wilson, E. O. (1998) *Consilience*, Alfred A. Knopf, New York, 367 pp.

7 Everything You Always Wanted to Know about Taxonomy but Were Afraid to Ask

Historian of science Mary Winsor took note of the improbable source of our knowledge of the diversity of life:

> What I have never directly confronted was the issue of how a modest number of men and women, a few of them clever or heroic but most of them quite ordinary, and several of them displaying serious weaknesses of character or insight, how this odd collection of people could have combined their efforts in such a way as to construct the enormous body of knowledge about life's diversity that is our heritage today. Whatever its present imperfections, systematics has certainly progressed.

A confluence of thought traditions and apparent contradictions, systematics is an enigma to many. While incorporating the latest theoretical and methodological developments, it is grounded in ideas, collections and traditions stretching back centuries. Illustrations of morphology bridge art and science. The study of phylogenetic relationships operates where history and science overlap. Non-experimental methods find justification in the philosophy of science. And, the desire to conserve as many species as possible, while documenting them all, straddles ethics and science.

Systematics is the foundation for understanding evolution, having created the predicate for the theory, yet it studies historical patterns, not evolutionary processes. It describes species but is not strictly descriptive, formal description incorporating explicitly testable hypotheses. It is exploration, research and scholarship all at the same time. It has taught us more about the diversity and origins of life than any other field, primarily by studying dead specimens. Its revisions and monographs focus narrowly on one taxon at a time yet are conceptually broad, spanning multiple continents and ecosystems through millions of years. It combines the adventure of exploration with the satisfaction of deep understanding and bridges the depths of time with the immediacy of discovery. It blends a love for species with rigorous scholarship, insatiable curiosity and passionate creativity. Progress depends on a community of scientists, yet demands the dedication of intense, prolonged individual effort. All of this is carried out with humility, recognizing

DOI: 10.4324/9781003389071-8

that a lifetime of work forges but one link in a long chain of scholarship. Intensely personal, taxonomic knowledge is nonetheless built, brick by brick, over hundreds of years by generations of experts. Individual contributions, no matter how large or impressive, are unmistakably part of an intellectual enterprise greater than self. Species names are as closely associated with individual taxonomists as paintings are with artists, yet they are intellectual property of humanity. Systematics is an individual and team sport at the same time. A community effort is required to create and test knowledge, and to assemble and curate research collections. Hypotheses about species and relationships are incrementally improved over generations as classifications are repeatedly revisited and improved. At the same time, many of the greatest contributions are made by inspired individuals working in near complete isolation.

As Benjamin Brewster wrote in the 1882 *Yale Literary Magazine*, "in theory there is no difference between theory and practice, while in practice there is." Defined by its usefulness to other disciplines, systematics is robbed of its most exciting, unpredictable and personally rewarding aspects. Like science in general, taxonomy is more complex and nuanced when real-world variables and the human element are introduced into its practice. Popular culture and scientists themselves have cultivated an image of scientists as detached, rational and deadly serious. Like characters in a low-budget 1950s sci-fi film, scientists are depicted as bearded old men in white lab coats, oblivious to an attractive female assistant making goo-goo eyes at the plot's hero. As a rule, this Joe Friday, just-the-facts-mam, characterization is wholly fictional. With thankfully few exceptions such inhuman, Spock-like scientists don't exist. It does call to mind one brilliant colleague, with near encyclopedic knowledge of his area of expertise, who is unbearably awkward in personal interactions. Every answer he gives is terse, usually a simple yes or no followed by an unnaturally long, intensely awkward silent stare. I admire his mind and value his contributions, but shipwrecked, stranded alone with him on a tropical island, would soon make me regret surviving.

Science can and should be enjoyable. I, for one, would not have dedicated my life to taxonomy if it were not fantastically exciting, deeply rewarding ... and a hell of a lot of fun. Maybe scientists fear that their ideas won't be taken seriously unless they are seen to be serious, too. Perhaps they are afraid that their employer will question their work ethic if they are seen to be having too much fun. Or, they may be concerned that the public, made aware of their human frailties, will no longer keep science in an exalted position among human pursuits.

My role models were so immersed in their research, so in love with the species they studied, and so engrossed in the ideas they pursued, that their profession was inseparable from their identity. Compulsively working evenings and weekends, rarely divorcing vacation from collecting trip, they put in ridiculously long hours enjoying every minute of it. They were appreciative of the privilege of being paid to do something they would have gladly done for free. I can almost hear their bosses saying, "Now, you tell me."

What makes science unique among human endeavors is not immunity from bias or blunder. Quite the opposite. Science is special because it is self-correcting.

It eventually arrives at, or at least closely approximates, truth, in spite of human shortcomings. The line between science and non-science, what philosophers call the demarcation principle, is deceptively simple: testability. Ideas, predictions or generalizations about the world that can be objectively tested by further observations or experiments are science. It doesn't even matter where ideas come from. They can be well supported by preliminary data, half-baked conjectures or wild-eyed speculations so long as they are framed in such a way as to be objectively testable. Unfounded and poorly conceived ideas, most with little or no supporting evidence at the outset, will, of course, be outed quickly when put to the test. Science, however, has nothing to say about the truth or falsity of untestable assertions. The claim that God exists, for example, is not scientific. It may or may not be true. Science is mute on the subject. This claim simply can't be tested and is, therefore, a matter of faith and a question for theology and metaphysics, not science.

This chapter shares a few of the joys and rewards in the life of a systematist, many of which lie well outside the stereotypic confines of science. I have never apologized for having fun, or for being all too human. It is absurd to perpetuate the myth that scientists are purely objective when I have yet to meet one who was. To the contrary, the very best science is inspired by strong emotions, not detached from them. In descriptive taxonomy, this is usually a love for the species themselves. Hotly debated theoretical ideas in systematics are another example. Conceptual advances may be motivated by the desire to demonstrate that a newly hatched idea is better than a prevailing one, or, just as often, to show that some other bastard's ideas are wrong. It really doesn't matter which, only that the idea is testable.

Reputations rise and fall by winning or losing debates, so it is no surprise when disagreements become impassioned and loud. While fights can become bitterly unpleasant, they need not. In fact, approached with the right attitude they are quite enjoyable. I have had a running argument regarding the definition of species with Professor Brent Mishler of the University of California at Berkeley for decades. Once, we were both giving lectures in a symposium on species at an international meeting in Canada. This was in the early years of Saturday Night Live, when Jane Curtin had popularized a particular phrase. My talk was first, followed by Dr. Mishler's. I said my piece and sat down. He took the podium, turned to look me straight in the eye and opened his speech with "Quentin, you ignorant slut." The Americans in the audience were bent over laughing as European colleagues frantically thumbed through bilingual pocket dictionaries. We sometimes go for years between seeing one another, then pick up our arguments more or less where they left off. After fighting like cats and dogs, genuinely at odds and with voices raised, we go out to share a few pints. That, to me, is science at its best. A clash of closely held ideas elevated above mean-spirited *ad hominem* attacks. I can, and do, strongly disagree with Dr. Mishler's ideas about species. Yet, at the same time, I respect Brent and, even more, consider him a friend.

Of course, Brent's pranks did not go unanswered. When he was on a study leave in Egypt, I needed to reach him regarding his chapter in a book I was editing. The only number I had for him was a fax machine in the office of the vice president of

Suez Canal University in Ismailia. I had a proper salutation on the cover sheet. But on the second page, the actual letter, I began simply "Dear Butthead," never mentioning Dr. Mishler by name. By the time my fax reached him at a remote site in the desert, hand delivered from the vice president's office, the cover sheet had been lost and the courier handed him the letter saying, "We thought it had to be you, the only American visiting."

Unfortunately, not everyone has Dr. Mishler's good sense of humor, nor the ability to separate ideas from the persons holding them. When I taught systematics theory, the inviolate rule in my class was that ideas may be stupid, but the individuals espousing them are not. In my experience, this is usually so, and, in any case, it makes for more productive discourse than calling someone a moron at the outset. I have always found such clashes of ideas invigorating and constructive, forcing me to either change my thinking or clarify the justification for my position. I cringe when I read in a newspaper that climate change, or any other matter, is "settled science." That phrase is the definition of an oxymoron. There is no such thing. If it's settled, it isn't science. And if it's science, it isn't settled. To be testable is to be open always to potential falsification. I do not subscribe to the idea of gravity because it is settled science; I accept gravity because planets stay in their orbits as predicted, and apples continue to fall on the heads of philosophers. There remains a chance, however slight, that things could be distressingly different tomorrow. Bearing this in mind keeps science honest and scientists on their toes.

The 1970s was an especially acrimonious decade for systematics. Tempers flared and insults flew because so many important issues and reputations were at stake. It called for thick skin and a sense of humor. Inside the tumultuous times, new ideas flourished. Hennig's then recent theories and methods were being explored, refined and extended, resulting in the greatest upheaval in how taxonomy is done in centuries. Outside the community, taxonomists earned a reputation of viciousness and incivility. But, how could it have been otherwise? The foundations of a science were being rewritten. It was taxonomy's version of Darwin's theory of natural selection or Copernicus' heliocentric conception of the solar system. A paradigm that had dominated the discipline since the 1940s was not only being challenged, it was being replaced. Advances were captured in Nelson and Platnick's "blue book," but if you want to examine some of the battle scars up close, you can do no better than exchanges in the back pages of *Systematic Zoology*.[1]

When I arrived at Cornell as a newly minted PhD, a retired faculty member offered me his early years of *Systematic Zoology*. I was thrilled to add them to my library but wondered why he would part with them. He explained that he had lost interest when the journal was turned over to theoretical arguments ... just the stuff that I loved. He was, I gathered, of the opinion that systematics is something you do, not something you think about.

In an academic twist on the circle of life, I have recently retired and dropped my subscription, too. The journal has changed its focus again, increasingly toward molecular techniques. With few papers relevant to comparative morphology or principles of classification, the journal has left me, just as it had my elder colleague. Although my personal journey from young Turk to old fart is apparently complete,

I remain convinced of the merits of phylogenetic theory and the pursuit of all informative evidence. If anything, morphology is more relevant and important today than ever. I worry that a disproportionate emphasis on DNA will discourage students who fall in love with a group of plants or animals from a career in systematics. It is fine to use DNA, of course. Even more, it would be just as wrongheaded to ignore molecular data as it is to neglect morphology.

My interests, like those of generations of taxonomists before me, are centered on species themselves. Understanding what makes each unique and how they are related, with a strong emphasis on morphology, natural history and classification. There are too many papers robotically using sequence data and off-the-shelf programs to assess relationships while never thinking deeply about individual characters. Ironically, these authors, in spite of the latest instrumentation and procedures, are also doing rather than thinking about systematics.

Systematics should welcome all evidence relevant to understanding characters, species and their relationships. Each data source tells us something new, serves as a test of other data, or both. While all data should be incorporated into the mix, it need not all come from single investigators or at the same time. It is almost inconceivable that any one person could have the knowledge and experience necessary to be equally good at digging fossils, sequencing DNA, mapping embryonic tissue differentiation and comparing morphology. For most mortals, specialization is indicated in the pursuit of excellence.

The assumption that DNA is superior data is scientifically unsound and fails to tap the passion and talents of those whose curiosity pulls them in other directions. The future of systematics lies in specialization, collaboration and synthesis, not conformity. Combining different sources of data guards against being misled by any single one. DNA tells us certain things, within a narrow band, with great precision, but misses a central point of systematics to understand other attributes of species.

There are so many pleasures, challenges and intellectual rewards in the life of a taxonomist that it is difficult to sum them up. Rather than attempting to do so, I will instead share a few that have been especially meaningful to me. Because of the diversity of both species and personalities, I suspect that each systematist has a list of her or his own. Some taxonomists have the collecting equivalent of a green thumb. Spending the same amount of time, in the same location, using the same collecting equipment they come up with more, and more varied, species than others. In contrast, there are specialists with little interest in, or aptitude for, field work. Years ago, there was a graduate student studying flies at Ohio State who worked exclusively with museum specimens. His approach was so extreme, faculty joked that he would not recognize a fly without a pin stuck through it. This leads me to the first virtue of systematics.

Playing to Your Strengths

Taxonomy includes a wide range of activities in the field, museum, library and laboratory. As a result, the number of species described or revised by a taxonomist each year can vary with the amount of support available. At whatever pace, the growth of knowledge is an ongoing process measurable in discoveries and collections. The Museum National d'Histoire Naturelle, Paris, not formally established until 1793, grew out of the royal garden of medicinal plants begun in 1635. It has since grown into one of the largest, most significant collections in the world and a site of major conceptual advances in systematics and evolutionary theory.

Depending on your focus and aptitudes, a career in taxonomy can be tailored to take advantage of your strengths, avoiding weaknesses and things that you would prefer not to do. Limitations of specialization are offset by community. Collaborations, whether in real time, or across generations, assures that all bases are covered while taking advantage of the special talents of each participant.

I have a particular passion for morphology. I find improbably complex and diverse structures and their origins endlessly fascinating. It is intellectually rewarding to begin with a jumble of different-looking structures and make sense of them in an evolutionary-historical context. In contrast, I personally find analyzing the sequence of nucleobases in a strand of DNA boring. I don't enjoy the kind of laboratory work involved and I'm uninspired by the simplicity of the evidence. To me, it may as well be chemistry and there is a reason that I chose to be a biologist. I am delighted that there are researchers who revel in molecular work, appreciate their efforts and applaud their contributions. The current preference for molecular data, however, is limiting our understanding of biodiversity. In the end, it is important that every informative source of evidence be explored by someone and integrated into our analyses and classifications.

It is human nature to do our best work when we enjoy the task at hand. I could sequence DNA if you held a gun to my head, but very few things short of that would entice me to do so. Disliking the process, and finding the data too removed from the characters I care about, I would never be driven to go above or beyond workaday quality. I prefer to leave DNA to those who will give it their best. It results in better data and makes my day a little happier.

I once reviewed a manuscript by a weevil specialist who was one of the best collectors I have ever known. As is often said, it is publish-or-parish in academia. His morphological drawings looked as if they had been done by a child with a blunt crayon. This did not reflect his knowledge of weevils, which was world class. He simply had no artistic talent and I must believe his sketches were drawn under duress. Science would have been better served allowing him to spend his time collecting and discovering new species rather than rendering poor drawings of little use. An ideal solution would have paired him with an artist. But taxonomy then, as now, was grossly underfunded.

For some systematists passion verges on obsession. I have always envied taxonomists with the discipline to sit at the bench and grind out species descriptions like Henry Ford manufactured cars. My intellectual curiosity too often sent me chasing tangents to count myself in their company. Norman Platnick of the American Museum of Natural History named more than 2,000 species of spiders. Dwight DeLong of The Ohio State University about 5,000 species of leafhoppers. On average, Charles Alexander, of the University of Massachusetts, named one new species of crane fly for each day of his very long career, more than 11,000 in all. And Edward Meyrick, an English schoolmaster and amateur entomologist, named about 20,000 species of Lepidoptera. Such massive numbers are not typical, of course, but overall averages could be much higher were taxonomists permitted to pursue their passion full time and given a modest number of support staff.

Taxonomy needs collectors, preparators, curators, informatics specialists, illustrators, cartographers, nomenclaturists, phylogenetic analysts, lab technicians, library researchers, copy editors, microscopists, DNA sequencers and fossil hunters, among others. It sits at a busy intersection of field biology, theoretical biology, geology, natural history, philosophy, history, geography, information science and art. Synthesizing diverse data and ideas is integral to taxonomy. Leading taxonomists must master many skills to some degree, know their limitations and be familiar with two centuries of literature in their group. But there is much to say for teamwork and delegation if we wish to accelerate species discovery and naming. A few renaissance systematists achieve excellence more or less across the board, somehow finding time to seemingly do it all. For most of us the quality and number of activities vary and that's okay, too. Science is best served by a community of specialists, each capitalizing on their knowledge, skills and proclivities.

Theoretical Systematics

Science is a restless business. Even with a well-established theoretical framework in place, there is always room for improvement and an ever-present possibility of an unexpected breakthrough. A recent revolution, initiated by Willi Hennig in the 1960s, transformed taxonomy and is not yet complete. Stopped in its tracks, distracted by technological advances, further conceptual and methodological progress awaits. As valuable as technology is in making evidence available, it will never replace theoretical clarity for understanding biodiversity or diminish the usefulness of classifications as the general reference system for biology.

Sciences go through cycles. Times of rapid theoretical or technological change are followed by periods of rote data gathering. As a graduate student, I came of age during the theoretical upheaval of the 1970s. So, I find mindlessly cranking out tons of data pretty boring. Bugs in a blender cannot compare to the intellectual joys of coming to know a group of species, and their complex characters, in rich detail and historical context. Had Hennig's revolution not been interrupted, had we maintained a balanced approach to evidence, it would be a good thing to focus on data gathering since species are going extinct so fast. As things are, we are in danger not only of losing the majority of species, but also of being left with empty museum drawers and unanswered questions about morphology, natural history and evolution.

The fact is, there remains important theoretical work to be done, not the least of which is employing molecular genetics to bridge genome and what morphology looks like. We need a deep dive into details of homology. This, however, requires that attention be returned to species and characters themselves, not just mass data gathering. Hennig made clear that what matters most is how data is analyzed, not where it comes from or how much of it exists. Morphological or molecular, fossil or developmental, the underlying theory sets the bar for quality of taxonomic work and reliability of conclusions. As with every complex challenge, this is not an either-or situation. We need to follow the phylogenetic revolution to its conclusion, improve technologies that facilitate data gathering and analysis and discover and describe as many species as we can. Because these endeavors enable one another, support to each benefits all.

Beauty, Awe and Wonder

Many scientists assiduously distance themselves from the arts and humanities. Because taxonomists use drawings and photographs to enhance the description of morphology, a connection to the visual arts is unavoidable. Moreover, for those willing to see it, beauty is everywhere in the structure of plants and animals. Some species are beautiful as a whole, such as a dahlia or dolphin. Others, admittedly less so. I have seen parasitic worms, molds and mites that only a mother could love. Even then, upon close examination, there are almost invariably aesthetically pleasing details of one sort or another.

I have always stood in awe before morphological diversity, innovative adaptations and the remarkable complexity of species. This appreciation was instilled in me as a child when I became fascinated by protists. The beauty of the single-celled ciliate *Paramecium caudatum*, with its complex organelles, is as difficult to deny as the elegance of an uncoiling *Vorticella campanula* or tumbling colony of *Volvox globator*. If you are not seeing beauty in nature, you are not looking closely enough. For self-evident reasons, I do not have a similar emotional response to DNA sequences that, to my eye, are more graffiti than art.

Every species has a beauty of its own, some fascinating structure, behavior or aspect of its evolution. The common name "slime mold" makes Myxomycetes sound repulsive, but their life cycle is nothing short of amazing. Mature fruiting bodies are as delicate and intricately pleasing in form, texture and color as flowers. When spores germinate, amoeba-like cells crawl across logs and leaf litter. At some point, initiated by cues not yet fully explained, these myxamoebae, as they are called, coalesce into one giant cell. Bordered by a single cell membrane, often containing thousands of nuclei, these include the largest cells on earth measuring up to four feet in diameter! The metamorphosis that follows is as impressive as that of caterpillar to butterfly. A gelatinous mass wells up to produce exquisite sporocarps that release spores, spread by wind and insects, to start the cycle over. An article by William Crowder in a 1926 issue of *National Geographic* is filled with paintings of slime-mold fruiting bodies and tales of a club of slime-mold collectors from New York City that made forays into the New England countryside in search of these rare beauties.

Admiration for the beauty of species is not limited to taxonomists and eccentric collectors. Visit an art museum and you will be emotionally moved by beautiful species captured in still life, landscape and *trompe-l'œil* paintings. High praise in poetry, song lyrics and literature often involve a comparison with beauty found in other species, which is why Walt Whitman said that a morning-glory at his window satisfied him more than the metaphysics of books.

Taxonomy is also tied to the humanities. Most obviously, systematics is a form of history. Not human history in the usual sense, systematics is nonetheless concerned with the story of humans and their relatives. Systematists share some kinds of questions, challenges and methods with historians of human events but have a fantastic advantage because evidence is recorded in genes and exquisitely complex morphological characters. Historians must get the chronology of events right before speculating about causes. If you have not established that the American Revolutionary War preceded the War of 1812 and followed the French and Indian War, you will be hard pressed to make sense of social and political issues precipitating these conflicts.

Biologists often speak of the "process" of evolution, but this is incorrect. There are many processes of evolution: gene mutations and drift, sexual selection, interspecies competition, as well as freak occurrences, such as a volcanic eruption randomly wiping out a nearby population. It is impossible to say with certainty which or how many processes contributed to the origin of any particular species. Before conjecturing, once again we must see events in their correct sequence. We must learn the chronology of appearance of characters and species. Some causal factors are precluded by the timeline. Founders of future species on remote islands could not have predated the islands themselves. And estimated absolute times, established by stratigraphy or molecular clocks, may suggest correlations of speciation with particular geologic events but are not infallible.

Phylogenetic analysis methods developed by systematists have been used to study other kinds of history, too, such as the sequence of hand-transcribed manuscripts and families of languages. Being non-experimental, taxonomists have been forced to explain and defend their science in an atmosphere that treats experimentation as if it were *the* scientific method. This has led to in-depth considerations of the philosophy of science. Epistemology, the branch of philosophy that explains how we know what we know, allows us to construct a firewall between science and non-science, without the need for experimental prescriptions or statistical measures of confidence.

Rules for naming species demand that systematists take account of every species named since 1758, making it historical in the dimension of human thought, too. It helps to consider also the social *milieu* in which taxonomy was done. To interpret 18th-century writings, we must recall the general acceptance of Creationism at the time. Patterns of similarity among species are objectively the same regardless of background assumptions, but interpretations vary with belief systems.

Systematics, along with aspects of geology, cosmology, paleontology and a few other fields, are remarkable for being both science and history. We think of the word history primarily in the context of human events, but the Universe and life have histories, too. Systematics is, by virtue of the conservation of heritable attributes, the most rigorous form of history. Human historians can only dream of so much "recorded" evidence. Using multiple lines of data, from morphology to fossils and molecules, taxonomists have a wealth of clues with which to piece together the chronology of species and their attributes. Anthropology, of course, is where early human history and systematics overlap broadly.

Intellectual Intimacy

It was my first trip to the Museum National d'Histoire Naturelle in Paris. In my left hand I held a *Hylecoetus* beetle described as a new species by the French coleopterist Léon Marc Herminie Fairmaire. On the desk was a photocopy of his description of the specimen published in 1889. Alternating my gaze between the microscope and his words, I struggled to understand what he was telling me about this specimen's morphology. He and I were two of fewer than half a dozen scientists who had ever examined this specimen closely. We were, in the moment, intellectually connected, sharing a morphology "idea space," in spite of being separated by time, culture, language and death. I concluded that Fairmaire was mistaken about the significance of this specimen. I, of course, had the advantage of examining hundreds of specimens collected after Fairmaire's time. This specimen merely represented a variation within *Hylecoetus dermestoides*, named by Linnaeus in 1761, but I now understood the thought process by which Fairmaire had been led astray.

What made this intimate meeting of minds possible was the museum specimen. I was holding the actual, individual beetle that Fairmaire had held in his hand so many years before. Like a wrinkle in the spacetime continuum, this specimen was a physical connection between he and I. A constant in an otherwise changed world. Because of the detailed observations involved, the connection became personal as I got into his head to understand his thoughts. Like Isaac Newton, every honest scientist admits that they stand on the shoulders of giants who came before. But only in taxonomy is this recognition canonized in nomenclatural rules and museum specimens.

Perhaps a better way to explain this phenomenon is by analogy with other physical connections transcending time. When my daughter Olivia wanted to visit the famous Civil War battlefield, we prepared by watching the film *Gettysburg*, then traveled to Pennsylvania. As we approached, the flat to rolling land we saw looked identical to fields we had been seeing for miles. But stepping out of the car, setting foot on that blood-soaked soil, is an experience unexplained by geography or landscape. Subconscious knowledge that you are touching the exact place where thousands suffered in anguish over three horrific days in July, 1863, is a wholly different experience than a photo, web site or history book.

This is why natural history museums will remain irreplaceable. Technically, you can see the bones of a tyrannosaur in photographs, drawings or manufactured replicas. But being in the presence of an actual specimen that lived in the late Cretaceous is an experience different in kind. Years ago, Dr. Timothy McCabe at the New York State Museum in Albany, mounted an impressive exhibit of moths he had collected in the New Jersey Pine Barrens, along with his collecting equipment. At the same time, the museum was renting a touring display of life-sized, growling animatronic dinosaurs. I was delighted to see that the line waiting to see Dr. McCabe's moths was longer than the queue for admission to the fake dinosaur show. People intuitively sense the difference and are drawn to the experience of being in the presence of the real thing, something museums are uniquely suited to deliver.

Science *con amore*

When I lived in London, there was a Raphael exhibition with paintings borrowed from as far away as South America. Arranged in chronological order, it was fascinating to see his artistic maturation. He was gifted as a teenager, of course. But when he traveled to Florence in 1504 and was exposed to work by da Vinci and Michelangelo, his paintings seemed to instantly transform from impressive to awe-inspiring. The best systematics, like great creations of art or literature, is inspired and executed with passion. Specialized knowledge, finely developed skills and experience take descriptive taxonomy—or art—only so far. Combined with genuine love for species themselves, a higher level of excellence is achievable and science is the better for it.

Many scientists arrogantly look down upon amateurs, mistaking their efforts as necessarily inferior. I prefer to think of all the best taxonomy as amateur—as opposed to amateurish—because excellence is neither the result of, nor necessarily correlated with, monetary compensation. After all, the word amateur refers to something done for love. If anything, one should be suspicious of professionals who risk, in being paid (or chasing grants!), becoming mercenaries. Looking at fads in science, often driven by funding trends and adulation, rather than the pursuit of knowledge for its own sake, one concludes that a mercenary mindset is widespread at both the individual and institutional levels.

A wise mentor taught me that you do not judge the quality of a university graduate program by its best student, but by its worst. Any program, even a bad one, can luck out with a self-made star student. When the student at the bottom of the class is achieving at a high level, however, you know the program is good. I suppose amateur took on a negative connotation because we chose to evaluate it by its worst performers. Yes, a larger proportion of unpaid, inadequately trained participants will achieve a lower level of excellence than paid ones. After all, it would be difficult to secure or hold a paid position unless you met certain minimum standards. All the same, the very best descriptive work is achieved, paid or not, when it is done *con amore*.

Creative Scientific Writing

There exists a sterile, formulaic approach to writing scientific papers that includes an introduction, materials and methods, results and discussion. In this standardized approach to writing, the colder and more detached the prose, the better. Papers are supposedly more professional when written in purely technical terms, cleansed of any trace of the humanity in the author.

For one member of the committee guiding my doctoral research, English was a second language. His command of the rules for English was impressively encyclopedic, but his strict adherence to them led to occasional disagreements. I recall a lengthy argument in his office over one sentence in my dissertation. I had written that an attribute of a particular species is "not uncommon." The rules of English, interpreted literally, say that this is a double negative. As such, he insisted that I rephrase it to say more simply that the attribute is common. But it is not, and I would not. There is a nuanced area in speech allowing for something to be neither common nor uncommon, to be in between, to be not uncommon. I am sure that was not my only literary offense in his eyes. I have been criticized for using flowery language on occasion, but to me this is the essence of human communication. I want to share my thoughts and feelings about the wonder and awe of nature, not simply record data. Such literary indulgences, to some, are evidence of taxonomy being a lesser science. They are wrong.

I prefer reading articles by passionate scientists who have command of the English language, or whatever language they are writing in, and are not afraid to use it, authors who do not hide their enthusiasm for purposes of feigned objectivity. I will take a passage like the one below by the late, great ground beetle expert George Ball any day over a dry technical report:

> A clear morning in February, 1966 found us collecting in a small patch of cloud forest on the northern slopes of a high mountain near Motozintli, Chiapas. The carabid fauna proved to be magnificently diverse, and the bromeliads yielded an especially rich harvest of specimens varying in shape, size, and color.

Had Ball simply listed the species he collected from bromeliads in Chiapas, I would have had the essential information. But I would not have been given a glimpse into his experiences as an explorer or felt so compelled to pack my own bags and head South.

Intellectual Excitement

It is difficult to explain the intellectual excitement of systematics research. The more you learn about a group of species, the greater and more frequent breakthroughs in understanding. These leaps of knowledge occur at multiple levels: individual characters, species and species relationships. Together, they lead the taxonomist on a fantastic intellectual journey, stretching the imagination with nearly unbelievable morphology, specimens from across the globe and visualizations of events played out over millions of years. What at first appears disjointed comes together in the context of history.

The male genitalia of ship-timber beetles, family Lymexylidae, consist of a central tubular structure, the median lobe, flanked by a pair of fingerlike lateral lobes. The lateral lobes in the genus *Atractocerus* are disproportionately large, exaggerated antler-like structures. When I first dissected a male of the genus *Melittomma*, I saw what appeared to be similarly bizarre lateral lobes. I soon realized, however, that diminutive lateral lobes were present on the genitalia separate from the much larger, antler-like structures that had caught my eye. Closer examination revealed that these structures were highly modified lateral elements of an abdominal segment, similar in size and shape (and presumably function, grasping the female during copulation), but sourced from entirely different body parts. Such examples of convergent evolution never fail to amaze me.

Another example is rhinoceros-like horns on the mandibles of two separate groups of species in the slime-mold feeding beetle genus *Agathidium* (family Leiodidae). Although superficially similar, occurring only on the left male mandible in both cases, these horns have evolved twice. In one group of species, it is the tip of the mandible that is prolonged and recurved to form a horn. In the other, the horn is a *de novo* structure arising from the dorsal surface about midway down the mandible.

These horns are not merely decorative. Kelly Miller collected *Agathidium pulchrum* from slime-molds in the Rocky Mountains and kept them for observation in our lab. He observed that when one male encountered another, he positioned his horn under the opponent, forcefully jerked his head upward and sent the competitor flying like a tiddlywink. For younger readers, Google it. Males also positioned their horns under females, but then gently bobbed their heads up and down, presumably doing something titillating for her that I leave to your imagination.

Attempting to comprehend the diversity among species, the origin and function of body parts, and the chronology of a multi-million-year history is not so straightforward as an experiment designed and executed. I agree with Crowson who said, "I think the imagination, properly disciplined by respect for evidence, has a very important part to play."

Adventure of Exploration

All taxonomists have stories of adventure in the course of field work. A few years ago, Andrew Hamilton and I organized a symposium at Arizona State University inviting systematists to share a few of their tales. Moderated by NPR journalist Robert Krulwich, the symposium presented one hair-raising tale after the next. One involved being stranded for days with no provisions atop a tepui in South America, until the weather cleared enough for a rescue helicopter to land. Another taxonomist was swept away in a flash flood, dashed against boulders, severely wounded and forced to crawl for miles to safety. Yet another, while chasing an insect, bumped a metal-handled collecting net into a high-voltage electrical wire. As you can imagine, it didn't end well and he was left with a smoldering hole where the toe of his boot had been. Joseph McHugh recounted the harrowing details of a trip he and I made that included being robbed twice and outrunning muggers with handguns in Montego Bay, Jamaica, and dining in the dark with newly made friends in Lima, Peru, after a power plant was bombed by the Sendero Luminoso. One cannot travel to remote areas of the globe and return without some good stories, assuming you survive to tell them.

Even less dangerous experiences can scar you for life. On that same trip to Peru, we stayed at a remote lodge on the Rio Tambopata in the sweltering Amazon where, in spite of no refrigeration, we were fed flesh from the same catfish for days. Delightful filets on day one gave way to soup boiled to kill bacteria a few days later. There was nothing else on the menu, so we choked down what we could. To this day, my stomach turns at the mere thought of catfish. On the bright side, I lost 20 pounds, learned that you can survive on a diet of bananas and the beetle collecting was spectacular.

Accountability and Recognition

The rules for naming species assure that taxonomists are never forgotten. Like an artist signing a painting, the person who names a species is permanently associated with it. Even in cases in which a species is no longer recognized as valid. This results in a deep sense of accountability and a strong incentive to do your best work. Unlike other fields of biology where a flawed publication is soon overtaken by more recent research and forgotten, species hypotheses, good or bad, are forever.

Mention Colonel Thomas Casey at any gathering of beetle specialists and you will hear snickering. His name has become a laughing stock, but his besmirched reputation is only partly deserved. There is such a thing as flat-out bad taxonomy. I once reviewed a manuscript reporting a remarkable new species in a genus I knew well. What made the species appear to be so unusual was that it had been misidentified. It was in fact a common species in a related genus. That's bad taxonomy. Casey's story is more nuanced.

Casey's first career was in the U.S. Army Corps of Engineers. After he retired from the Corps in 1912, he pursued his love of beetles with a vengeance, naming more than 9,000 species, subspecies and varieties, all from North America. Many of Casey's species are no longer considered to be valid. There is no better example than his ground beetles (family Carabidae). Of 1,376 Casey names for carabids (excluding tiger beetles, considered a separate family at the time), the imminent Swedish ground beetle specialist Carl Lindroth recognized only 81 as names for biologically valid species. Ouch!

What is unfair to Casey is that he was both a keen observer and knew his beetles well. He did not describe species in the wrong genera. His primary problem was his concept of species, which was a notch off of reality. He was the consummate splitter, naming minor variations without having studied enough specimens to realize that there were intermediate forms. This was perhaps foreseeable, given the number of his papers with some variation on the following title: "New species of [insert taxon name here] from my cabinet." He fell into the trap that many general biologists do, failing to distinguish between genetic traits that vary within a species and characters constantly distributed within and thereby demarcating species. Regardless, Casey, who is forever associated with 1,295 ground beetle names that do not correspond to actual species, now suffers a tarnished reputation.

It is said that a beetle collector, Henry Wickham, cashed in on Casey's tendency to name variants. Because Casey paid Wickham for each new species he collected, Wickham would intentionally throw away intermediate forms from a long series of beetles, deliberately misleading Casey into naming several species where there clearly was one, and making himself a little wealthier in the bargain. A case of Coleoptera capitalism at its worst.

Greater than Self

Whatever your passion in life, there is enormous satisfaction in the knowledge that your labors are contributing to something greater and more enduring than your mortal self. Nowhere is this more evident than in taxonomy where each scientist is a stepping stone in the progress of knowledge. When you become an expert on a group of species, you receive centuries of accumulated knowledge, museum specimens and classification schemes, with the challenge to pass them on in an improved and expanded form. This is both humbling and inspiring. Playing such a role in a centuries-long scientific drama is as close to immortality as you can get.

Joy, Humor and a Little Naughty

Because systematists love the organisms they study, the great joy in their work sometimes spills over in the form of masterful publications, beautiful illustrations and good humor. Examples of the latter can be found in titles of publications, commentaries and even species names. Jonathan Coddington's review of a book on sexual selection and male genitalia by William Eberhard was titled "The shape of things that come." And, in characteristically restrained British wit, Peter Barnard said in a review of a book on aquatic insects that it was not so much a second edition as it was a reprinting of the first edition with new errors introduced by the author.

There are a large number of humorous species names, too, among which I have a special fondness for puns, such as Terry Erwin's *Agra vation*, *Agra cadabra*, and *Agra phobia*; Norman Platnick's *Apopyllus now* and three genera related to, but not, *Nops*: *Notnops*, *Tisentnops* and *Taintnops*; Paul Marsh's *Heerz lukenatcha* and *Heerz tooya*; Neil Evenhuis's *Pieza kake*, *Pieza pi*, *Pieza rhea*, *Pieza deresistans* and *Riga toni*; Cornelius Philip's *Tabanus nippentucki* and *Tabanus rhizonshine*; Alan Solem's *Ba humbug*; Paul Spangler's *Ytu brutus*, that my high school Latin teacher would have appreciated; and a few that I had a hand in along with Kelly Miller: *Gelae baen*, *Gelae donut*, *Gelae fish* and *Gelae rol*. Feel free to groan.

At times, names take a naughty turn. I laughed out loud in 2009 when a paper by Adam Slipinski and colleagues created a new tribe of corylophid beetles, Foadiini, with a name based on the genus *Foadia*. I do not know whether they were aware of its true etymology, or not. When James Pakaluk named *Foadia*, for the record—and to get the name past the editor—he stated the root "f-o-a-d" was an arbitrary combination of letters, something perfectly acceptable under the zoological code of nomenclature. In actuality, f-o-a-d was an acronym for "*uck off and die." Jim, the first student in my lab at Cornell, tragically died young. He would have enjoyed the elevated stature of his naughty name.

As we reimagine systematics to confront the biodiversity crisis, it is important that we not lose sight of its many dimensions. Being able to identify species is a necessary goal that could, at least potentially, be realized by cold, impersonal molecular data, computer algorithms and unimaginative arbitrary names or, even worse, some other unique identifier, like a number or DNA sequence. But this would squeeze some of the best of humanity from the process of doing taxonomy. And rob us of so much joy, humor and creativity that it should not be seen as an acceptable substitute for taxonomy's traditions. We owe it to ourselves and future generations to retain the best of systematics as we inspire the next generation of species explorers to dream big, achieve much and enjoy themselves. But first, we must overcome a few misconceptions.

Note

1 The journal *Systematic Zoology* has since been renamed *Systematic Biology*. The opinion pages of the "Forum" section were particularly drenched in venom, blood and guts.

Further Reading

Ball, G. E. (1975) Pericaline Lebiini: Notes on classification, a synopsis of the New World genera, and a revision of the genus *Phloeoxena* Chaudoir (Coleoptera, Carabidae). *Quaestiones Entomologicae*, 11: 143–242.

Coddington, J. A. (1987) The shape of things that come. *Cladistics*, 3: 196–198.

Crowder, W. (1926) Marvels of Mycetozoa. *National Geographic*, 49: 421–443.

Crowson, R. (1982) Computers versus imagination in the reconstruction of phylogeny. In *Problems of Phylogenetic Reconstruction* (eds. A. Joysey and A. E. Friday), Academic Press, London, pp. 245–255.

Erwin, T. L. (1986) *Agra*, arboreal beetles of Neotropical forests: *Mixta* group, *virgate* group and *ohausi* group systematics (Carabidae). *Systematic Entomology*, 11: 293–316.

Erwin, T. L. (2002) The beetle family Carabidae of Costa Rica: Twenty-nine new species of *Agra* Fabricius 1801 (Coleoptera: Carabidae, Lebiini, Agrina). *Zootaxa*, 119: 1–68.

Evans, H. E. (1968) *Life on a Little-Known Planet*, E. P. Dutton & Company, New York, 318 pp.

Evenhuis, N. L. (2002) *Pieza*, a new genus of microbombyliids from the New World (Diptera: Mythicomyiidae). *Zootaxa*, 36: 1–28.

Hennig, W. (1966) *Phylogenetic Systematics*, University of Illinois Press, Urbana, 263 pp.

Lindroth, C. H. (1975) Designation of holotypes and lectotypes among ground beetles (Coleoptera, Carabidae) described by Thomas L. Casey. *Coleopterists Bulletin*, 29: 109–147.

Marsh, P. M. (1993) Descriptions of new Western Hemisphere genera of the subfamily Doryctinae (Hymenoptera: Braconidae). *Contributions of the American Entomological Institute*, 28: 1–58.

Mayr, E. (1969) *Principles of Systematic Zoology*, McGraw-Hill, New York, 428 pp.

Mayr, E. (1982) *The Growth of Biological Thought*. Harvard University Press, Cambridge, 974 pp.

Miller, K. B. and Wheeler, Q. (2004) Two new genera of Agathidiini from the Nearctic and Neotropical regions (Coleoptera: Leiodidae). *Coleopterists Bulletin*, 58: 466–487.

Nelson, G. and Platnick, N. (1981) *Systematics and Biogeography: Cladistics and Vicariance*, Columbia University Press, New York, 567 pp.

Ohl, M. (2018) *The Art of Naming*, MIT Press, Cambridge, 294 pp.

Padial, J. M., Castroviejo-Fisher, S., Köhler, J., et al. (2008) Deciphering the products of evolution at the species level: The need for an integrative taxonomy. *Zoologica Scripta*, 38: 431–447.

Pakaluk, J. (1985) New genus and species of Corylophidae (Coleoptera) from Florida, with a description of its larva. *Annals of the Entomological Society of America*, 78: 406–409.

Platnick, N. I. (1994) A review of the Chilean spiders of the family Caponiidae (Araneae, Hapogynae). *American Museum Novitates*, 3113: 1–12.

Platnick, N. I. and Shadab, M. U. (1984) A revision of the Neotropical spiders of the genus *Apopyllus* (Araneae, Gnaphosidae). *American Museum Novitates*, 2788: 1–9.

Rader, K. (2004) *Making Mice: Standardizing Animals for American Biomedical Research, 1900–1955*, Princeton University Press, Princeton, 312 pp.

Ross, H. H. (1974) *Biological Systematics*, Addison-Wesley, Reading, 345 pp.

Schuh, R. T. (2000) *Biological Systematics: Principles and Applications*, Cornell University Press, Ithaca, 236 pp.

Simpson, G. G. (1961) *Principles of Animal Taxonomy*, Columbia University Press, New York, 247 pp.

Slipinski, A., Tomaszewska, W., and Lawrence, J. F. (2009) Phylogeny and classification of Corylophidae (Coleoptera: Cucujoidea) with descriptions of new genera and larvae. *Systematic Entomology*, 34: 409–433.

Solem, A. (1982) Endodontoid land snails from Pacific Islands (Mollusca: Pulmonata: Sigmurethra). Part II. Families Punctidae and Charopidae, Zoogeography. Field Museum of Natural History, Chicago, 336 pp.

Spangler, P. J. (1980) A new species of *Ytu* from Brazil (Coleoptera: Torridinicolidae), *Coleopterists Bulletin*, 34: 145–158.

Wheeler, Q. D. (1986) Revision of the genera of Lymexylidae (Coleoptera: Cucujiformia). *Bulletin of the American Museum of Natural History*, 183: 113–210.

Williams, D. M. and Knapp, S. eds. (2010) *Beyond Cladistics: The Branching of a Paradigm*, University of California Press, Berkeley, 330 pp.

Winsor, M. P. (2001) The practitioner of science: Everyone her own historian. *Journal of the History of Biology*, 34: 229–245.

8 A Science Misunderstood Greatly

As a student at Ohio State, I took insect morphology and several advanced taxonomy classes from Professor Donald J. Borror. His general knowledge of insects was impressive, the product of intense curiosity and decades of refining his textbook. One sunny day he walked the class around Mirror Lake to observe dragonflies on the wing. Seemingly on cue, a large dragonfly zoomed by and a student asked, "Professor Borror, what was that dragonfly?"

"That was *Anax junius*."

"How could you tell?"

"It has a bullseye marking on its head."

The student looked perplexed as Don held back a smile.

As anyone who has examined *Anax junius* under a microscope can attest, they indeed have a bullseye-shaped mark on the head, just above the eyes. Don, of course, had spent countless hours observing dragonflies and knew all the species that might be seen on campus. So, its size and flight pattern were more than enough for positive identification.

To the uninitiated, taxonomists develop what may look like a sixth sense. After many years and thousands of observations, experts like Borror may come to recognize species without a clear view of confirmatory attributes. In a genus I studied, there were three closely related species that were challenging to distinguish. Early in my project, I had to scrutinize and sometimes dissect specimens before being certain of their identities. By the end, I had developed an "eye" for the species and, based on subtle, yet reliable, differences in the size and density of punctures on their dorsal surface, rapidly and accurately sort them. This reflected experience more than intuition. Until someone has examined a sufficiently large number of specimens, it is not an easily transferrable skill.

For a biologist trained to methodically follow, step by step, protocols for an experiment, this ability can look more like black magic than science. But such experiential honing of perception is no different than a seasoned radiologist who easily spots slight differences between x-rays that are invisible to the rest of us, and similar to a musician playing an instrument without thinking about it or a basketball player relying on muscle memory to make a three pointer. Practice, practice, practice.

DOI: 10.4324/9781003389071-9

Most biologists think in terms of how the world operates. At a cocktail party, they introduce themselves and ask, "What question do you study?" When I respond that I study beetles, not a question, the conversation, or at least the topic, suddenly ends. It is as if you are speaking a foreign language. To study processes the most rigorous approach is experimentation. So, this is the frame of reference for the majority of biologists. Failing to conform to this norm, systematics appears so different to them that they question whether it is really science. Considering that scientists are in the business of pushing the boundaries of knowledge, I am surprised by the prevalence of such slavish conformity. This explains, in part, a prejudice among natural scientists against the social sciences and humanities. Such intellectual xenophobia is a product of peer pressure and a failing of our education system.

The public harbors no such preconceptions and is receptive to stories of species discovery through wits and observation and, when it comes to field work, serendipity. But to restore the stature of descriptive systematics we must both overcome bias among biologists and build support beyond. With thousands of remarkable species discoveries each year, systematists have a wonderful opportunity to trumpet the news, polishing taxonomy's reputation. Cultivating public awareness can create allies for restoring support to systematics.

When I was at the NSF, a program in neurobiology was on the chopping block because it leaned a little too far toward biomedicine and away from basic science, indicating the NIH might be a better home for it. Affected researchers mobilized a letter writing campaign, and the program was quickly saved. Only about 8,000 letters were written, but they represented nearly every congressional district in the country. One predictable thing about politicians is that they go weak-kneed when large numbers of constituents speak up. When congressional aides begin calling an agency with concerns, the agency, totally dependent on Congress for its budget, reacts. The public, even in the form of a modest number of scientists, has a stronger voice than is generally appreciated.

As education requirements for biological subdisciplines have expanded, room in the curriculum for broader education has shrunk. There was a time when every biologist took at least one taxonomy course, but those classes have been squeezed out along with requirements in the humanities and social sciences. In extreme cases, students are being trained more than educated. Their skill set prepares them to carry out specialized work expertly but undermines their ability to critically question how they do things or see their work in its broader context.

As descriptive taxonomy was neglected in recent decades, technological innovations proceeded. The result is that almost everything that a traditional systematist does can be accomplished faster and more efficiently with no loss of quality. From travel to remote field sites to accessing museum specimens, capturing images of morphology, analyzing data and disseminating knowledge, there are off-the-shelf solutions. The one great remaining obstacle to realizing taxonomy's potential is changing minds.

It is imperative that everyone—professional biologists, the public and leaders of academic and research institutions—understands and appreciates what makes contemporary systematics rigorous, fascinating and relevant. The reliability and information content of scientific names at every level, from species to kingdom, depends on how well systematics is done—and how often its hypotheses are tested. The current push to rely on a single data source to both recognize species and assess their relationships is eroding the breadth and depth of systematics.

In *Self-Reliance*, Ralph Waldo Emerson asked,

> Is it so bad, then, to be misunderstood? Pythagoras was misunderstood, and Socrates, and Jesus, and Luther, and Copernicus, and Galileo, and Newton, and every pure and wise spirit that ever took flesh. To be great is to be misunderstood.

I hesitate to disagree with Emerson, but my observations of systematics suggest that being misunderstood isn't all that great. In the course of my career, I have watched as the mission of systematics was misinterpreted through ignorance or malice, hastening a decline in taxonomy's prestige, positions and funding. No science is more intellectually challenging or rewarding than systematics, none has a greater or more direct impact on science and society, yet no science is so widely misunderstood. Support for systematics dries up when its aims are misconstrued. Or, just as bad, support is limited to applied taxonomy, providing support to other sciences while neglecting fundamental research. With millions of species racing toward extinction, withdrawal of support from descriptive taxonomy is a scientific tragedy.

Some misunderstandings about taxonomy result from a lingering reputation no longer deserved. Yet, as Luc et al. observed

> Decision makers came to consider taxonomy as a science of the past, one that belongs to museums or that may even not be a science at all. Consequently, the word "taxonomy" gradually became a "dirty word," to the extent that any research work including it in its heading would most probably not be funded.

There was a time when deciding what was, or was not, a species, or how to group species in a classification, were nakedly subjective. Experts, with years of experience, developed intuitive rationale that, even though frequently right, were neither explicit nor objectively testable. It was the ideas and methods of German entomologist Willi Hennig that led to the replacement of speculation with critical hypothesis testing. Although he had refined his ideas for decades, it was the English language translation of *Phylogenetic Systematics* in 1966 that sparked a revolution.

In spite of the ensuing transformation, the impression that taxonomic decisions and classifications are arbitrary remains. To diminish the reputation of systematics today, because of limitations in its past, is like belittling astronomy because Ptolemy's geocentric model of our planetary system was once universally accepted. The theoretical breakthrough in systematics may be recent, but that makes it no less profound. Ironically, after centuries of support, now that systematics has joined the ranks of rigorous science, ready to fulfill its potential, its mission is ignored.

Early in the 19th century, when physiologists adopted the experimental method, it spread through biology like wildfire. Now, the experimental method is so ubiquitous among the life sciences that it is mistaken by many as a litmus test for science itself. So, let's be clear: all well-designed experiments are science, but not all science is experimental. Because students majoring in biology today are rarely required to take a single course in the theoretical foundations of systematics, or even the philosophy of science, it is understandable that they do not recognize advances in the field.

If you were indoctrinated with the experimental method as a student, you may question whether any non-experimental approach is legitimately science. Or, if you are a field biologist, you may see taxonomy as nothing more than a way to identify species and call them by name. But, as Bremer et al. said, "Taxonomy is not a service function for labeling organisms, but a science of its own, dealing with variation, relationships, and phylogeny."

Not only is taxonomy observational rather than experimental, it is engaged in the study of history, a subject generally thought of as a humanity, not a natural science. Justification for systematics-as-history being scientific, however, is found in the definition of science itself. As discussed elsewhere, the difference between science and non-science is not the experimental method, it is *testability*.

So, how does systematics as a rigorous, non-experimental, comparative, observational, historical science work? My favorite philosopher of science is Sir Karl Popper. I first studied his ideas as a student when I read *The Logic of Scientific Discovery*, a book my fellow graduate students and I affectionately referred to as LSD. True to the nickname, it altered my mind and I saw things I had never seen before. Popper is dismissed by many philosophers of science and experimentalists, and I have concluded why. Experiments have so many possible outcomes that statisticians speak of a "Universe" of outcomes and have, out of necessity, developed techniques for deciding whether a particular outcome is significant or due to chance. With such messy results, I see why Popper's philosophy is not an easy fit for experimental biology. But the messy state of affairs in experimentation should not constrain systematics that deals instead with less ambiguous, more elegant hypotheses.

Put simply, the more susceptible a claim about the world is to being shown false, the more rigorously scientific. In Popper's philosophy, nothing can be proven. Things can only be disproven or falsified. This is why his approach is often called falsificationism. A classic example of a Popperian statement is, "All swans are white." Regardless of how many white swans you observe, it is impossible to prove the statement correct. We can never knowingly see every swan. On the other hand, we need only observe a single black swan and the statement is refuted. Thus, the most rigorous of all scientific hypotheses are all-or-nothing claims about the world that leave no wiggle room. Generalizations that can be toppled by a single observation. This is precisely the kind of hypothesis that permeates systematics.

Every single individual of a species is claimed to share a unique, specified combination of heritable characters found in no other species. Similar all-or-nothing claims are made about groups of related species, too. All angiosperms bear flowers; all insects have six legs; and all vertebrate animals have a "backbone." No exceptions. It takes a single flowerless angiosperm, one insect with fewer than six legs, ignoring victims of sadistic school children, or a vertebrate animal without a segmented backbone to refute these generalizations. No scientific hypothesis is subjected to more strenuous, unforgiving testing than one that is so vulnerable that it can be rejected by a single observation.

It would be disingenuous to suggest that, while theoretically elegant, testing hypotheses is always as simple as looking at specimens. Characters used by systematists are theoretical constructs that may or may not align with visual first impressions. It requires serious intellectual work to establish, for example, that all reptiles are quadrupeds—animals possessing four legs. You may be thinking that, last time you looked, snakes didn't have legs at all, yet they are reptiles and as such quadrupeds. There is a fundamental difference, however, between primitively legless animals—like earthworms or fish—and those which have secondarily lost legs. The former can only be traced to ancestral species that also lacked legs; the latter to ancestors with legs. In the case of snakes, characters other than legs point to their being reptiles. Further, a careful examination reveals that they have both the genes coding for legs and evidence of tissue buds that are embryonic precursors of legs in all quadrupeds. Their "legs," understood to include genes and incomplete developmental pathways, are present but highly modified, never fully developing. Thus, the character "four legs" *is* shared by snakes, in spite of superficial appearances. This is why characters in systematics are theoretical constructs. And why taxonomic descriptions of species are not merely accounts of what individual specimens look like.

The combination of Popper's philosophy, Hennig's phylogenetic revolution and Linnaeus' ambition to inventory all species positions systematics for great success documenting, classifying and understanding the diversity of life. Darwin knew that a fascinating story remained to be told, spelling out how, exactly, the diversity of species came to be. That story is written in improbable, complex morphological attributes: the very properties of species being deemphasized by a growing reliance on molecular data.

In order to advance the mission of systematic biology, it is necessary that experimental biologists and the public be dissuaded from the belief that only experiments are reliable science, that classifications are necessarily subjective, that molecular data is superior to morphological and that species are arbitrary, carved out of more or less continuous genetic variation for convenience. It is curious that systematics is accused of being anachronistic by those clinging to such outdated beliefs.

It is important to debunk myths and misconceptions that haunt systematics. These are used as excuses to deny respect and funding to taxonomy by those ignorant of its contemporary theories and methods, unclear on its mission and goals and disinclined to share limited research dollars. As the biodiversity crisis rages, we waste valuable time by perpetuating such unwarranted assumptions and suspicions. So, let's debunk some myths.

Species Are Arbitrary

A surprising number of biologists do not believe in species. This is a legacy of the "population thinking" of the New Systematics of the 1940s or, in other instances, the misapplication of "phylogenetic thinking" at the population or even individual level. Buying into the myth that species are arbitrary fabrications, they conceive of life as a continuum of genetic variability. This confuses the genetic landscape within species with irreversible discontinuities that exist following the completion of speciation. Any seasoned field biologist can attest to the existence of species—of distinct *kinds* of living things. It is true that species are most easily recognized in the context of local habitats. And that, when all genetic variations, across all populations, of a geographically widespread species are considered, circumscribing a species can be challenging.

Poor assumptions beget poor results. One can make up nonsense species, such as people with red hair or populations separated by an arbitrary percentage of DNA, but this does not take away from the rigor of species that are properly formulated. As Kevin Nixon and I discussed, it is important to distinguish between informative, constantly present *characters* and variable *traits* that are, while useful to population studies, uninformative at the species level. Traits that vary within a species cannot be used in its characterization. Only "fixed" characters, those shared by each and every individual within the species, are suited to this end. It is intellectually dishonest to point to variable traits as evidence that uniformly shared characters, or the species possessing them, do not exist. The notion that species are arbitrary traces in part to the unfortunate practice of naming partially diverged populations as subspecies. Evolutionary taxonomists used to cite rules of thumb for subspecies, naming populations that were diverged by an arbitrary percentage, generally around 80%. In such cases, I agree that these "kinds" do not exist. These may be species in the making, but until they are 100% differentiated, indicated by character transformations, they have not achieved that status. And there is a chance that they never will. We can never observe every individual or prove that populations are truly completely diverged, of course, which is why species are hypotheses. Characters, not traits, are the evidence for species and clades.

Let's consider a character shared by Araneae. All spiders, and only spiders, possess silk-producing organs called spinnerets. Many spiders have six spinnerets, others two, four or eight. This variation in number, as well as size, shape and position relative to one another, does not change the fact of their ubiquitous presence in 100% of spiders—and their absence in every other living thing. Such diversity of form illustrates why the character "spinneret" is a theorical construct, not a description of what any particular spider's spinnerets look like.

The job of systematists is to discover and hypothesize the evidence for species and groups of related species, not to pursue evolutionary processes that may or may not have played a role in their origins. This involves a detailed assessment of the distribution of attributes that cannot be accomplished by experiments. It's fun to think about evolutionary processes, to imagine the forces driving speciation, but this is not the role of taxonomy and is, in any case, speculative. Taxonomy limits the number and kind of candidate processes to those consistent with the historical pattern but cannot unequivocally link one or more processes to any particular character transformation or speciation event. Admittedly frustrating, this is a limitation of studying history, not a shortcoming of systematics. It is the task of geneticists to sort out what processes are possible, under what circumstances, thus refining what remains conjecture when we attempt to associate processes with patterns from the past. Together, the patterns discovered by systematists, and candidate processes demonstrated by geneticists, create a powerful portfolio of ideas with which to comprehend how evolution works today and may have worked in the past.

To summarize, species properly formulated are not artificial constructs, but the result of careful analysis of the distribution of meticulously conceptualized characters. They are demarcated by a unique heritable attribute, or unique combination of heritable attributes, that is constantly distributed among all individuals within a species, and none beyond. Character hypotheses are independent of assumptions about evolutionary processes and thus compatible with all possible processes.

Systematics Is "Merely Descriptive"

One accusation is so often leveled at systematics that it deserves further comment. Obviously, it is true that taxonomists record what species look like. In a formal species description, one of two approaches is taken. It may consist of a detailed description of a single specimen followed by, or incorporating, a discussion of variation seen among individuals within the species. Or, it may be a composite description, a distillation of all specimens observed, encompassing a full account of the range of variation present within the species. Descriptions, diagnoses and associated discussions include the combination of characters unique to the species, and those shared with relatives. Information about variable traits is included not as evidence of species status, but to assist in making identifications and avoid confusion of the species with others which may be similar. The circumscription, the hypothesis and the idea of a species are based only on the combination of characters that, *as defined*, do not vary within the species. Writing reliable and accurate descriptions requires familiarity with systematics theory and sophisticated understanding of characters. Uninformed users of taxonomic literature may fail to recognize the inherent strengths and limitations of such theory-infused descriptions.

Systematics Is a Service

Systematists provide essential services by identifying species, making species identifiable, elucidating relationships and organizing species in phylogenetic classifications. This is why they have developed identification tools, from traditional dichotomous keys to interactive diagnostic programs (e.g., keys.lucidcentral.org/search/) and image recognition software (e.g., PlantSnap), some of which incorporate artificial intelligence so as to become better at identifications over time (e.g., SPIDA-Web). But do not confuse services with science. It is not coincidental that proponents of molecular-based taxonomy are concerned entirely with identifications on one hand and relationships among species on the other. What these two have in common is that they are useful to biologists outside taxonomy. But they ignore the fascinating stuff that exists between, the diversity of morphology that has driven systematics from its inception. Systematists formulate species hypotheses to get at and understand the kinds of organisms that exist. They name species so that information may be stored, retrieved and communicated. They analyze relationships so that the origins of diverse morphology may be understood. And they classify species to summarize and organize all that has been learned. But without morphological descriptions to flesh out the uniqueness of each species, limited to simply telling species apart and assessing their relationships, systematics is very much closer to being a mere service. Attention to the needs of general biologists explains why so many phylogenetic analyses are devoid of deep thought about individual morphological characters. As a science, rather than service, a focus on individual characters is as central to systematics as considerations of species and relationships among them. This is what advocates of a molecular-based taxonomy miss as they threaten the integrity and information content of taxonomy in order to service general biology.

DNA "Reveals" Phylogenetic Relationships

Too often, the results of DNA analyses are treated as if they were revelations of truth about phylogeny. Even when DNA data are parsed as discrete characters, rather than used to assess overall similarity, they are not infallible. Resolving conflicts between patterns indicated by DNA and morphology involves a critical reassessment of each with no presumption as to which, if either, has been correctly interpreted.

DNA Data Is Superior

As Ed Wilson said to me at a meeting in Washington, quietly and offstage, molecular data doesn't receive more money because it's better than morphology; DNA is assumed to be better because it receives more money. DNA's prestige is derived from a combination of impressive technology and riding on the shirttails of massively well-funded biomedical science. It is worth keeping in mind that systematics is primarily a historical science and DNA, by virtue of its simplicity, contains less historical information than morphology. Even more, the most striking evolutionary novelties cannot be recovered from DNA alone. Although it need not be so, the popularity of DNA has become a distraction from the mission of systematics.

The exaggerated stature of DNA is also due to a smoke-and-mirrors trick. When DNA upsets long-standing ideas about relationships among some group of plants or animals, it is appropriately hailed as exciting progress. But why? The only reason such new views of relationships are interesting is that they allow us to reinterpret a body of descriptive knowledge that already exists. If centuries of preceding descriptive taxonomy did not exist, there would be nothing of interest to reinterpret with DNA results. It is a slight of hand to celebrate insights from a DNA analysis while simply taking for granted the existence of descriptive work without which its results would be trivial.

Thus, even if you are a beady-eyed advocate of DNA analyses, you ought to be a forceful proponent of descriptive taxonomy. Unless existing hypotheses about morphology are tested and corroborated, and unless the morphology of millions of yet-to-be-discovered species are described, DNA will have precious little of interest to explain in the future. This may not worry those who simply want to identify species, but it is a grave concern to anyone interested in species, evolution or systematics as a science.

Anything an Amateur Can Do Can't Be Serious Science

Systematics includes many activities, some simple, others quite demanding. Collecting and preparing specimens requires training, but not a great deal of formal knowledge. Interpreting complex morphology depends on a mastery of empirical facts, theories and methods. Depending on particulars, amateurs can play important, hands-on roles in many aspects of taxonomy. This is a strength, not a weakness. The multiplier effect of an army of amateurs being places, making observations and collecting specimens that professionals do not have the numbers or time to do is an enormous, insufficiently tapped resource.

If we are smart, we will formalize and expand the role of amateurs in an inventory of species, with appropriate oversight. As a cyber-platform for taxonomy is implemented, it will become increasingly easy to train amateurs and to verify the quality of their work. Beyond important direct contributions, such as adding to known genetic and geographic variation, preparing specimens and entering data, their engagement cultivates a supportive citizenry capable of advocating for systematics, an inventory and natural history museums.

For a few highly motivated amateurs, it remains possible for them to become experts in their own right. As systematics has become more sophisticated in its theories and methods the bar has risen significantly, yet native intelligence, hard work and perseverance keep credible amateur taxonomic contributions attainable. Advances in digital imaging and telemicroscopy mean that amateurs can examine rare, delicate museum specimens from afar, removing a traditional barrier for non-professionals. Such access will continue to expand, inviting broader participation. This is all to the good, assuming that we can avoid a proliferation of amateurish, unprofessional contributions by using peer review, training and supervision.

Natural History Collections Are a Relic of a Bygone Age

As molecular studies occupy center stage, natural history collections are increasingly seen as anachronisms. Their care is entrusted to curators and collection managers who, although professionals, are paid to work independently from taxonomic research programs. This reflects a fundamental misunderstanding of why collections are valuable and how they are put to their highest use. There should be no daylight between taxonomic research and curation.

Collections must remain a primary research resource for systematic biology if it is to fulfill its mission. You can tell how healthy taxonomy is in a natural history museum by the activity in its collections. Is the museum growing and developing collections? Are researchers focused on the use of collections in revisions and monographs? Is curation up to date, reflecting the latest knowledge? Is there a revolving door of visiting taxonomists? The specimens they contain, and information extracted from and attached to them, represent the greatest amount of knowledge of biodiversity that exists. A healthy collection is a dynamic center of research and discovery, not a closet of artifacts to be preserved and rarely touched or examined.

Properly maintaining collections is a costly, demanding job requiring a team of collaborative expertise: taxon experts contributing to research and curation, preparators accessioning new material, collection managers overseeing care, growth and development, data managers and technicians, among others. Mass extinction indicates this should be a time of hyper-activity, unprecedented growth and constant use of collections along with visionary leadership from the top.

Specialization, even with good intentions, may go off the rails. Some museums have separated the functions of research and curation, which makes little sense. The two are interconnected: research conclusions are reflected in collection growth, development and arrangement, and this facilitates further research. The creation and use of knowledge must constantly flow between and bind the two. In the most extreme case, systematists are effectively banned from collections. They must request specimens for examination from collection managers who then retrieve them. While the two are partners with the best interest of collections at heart, they at times find themselves at cross purposes. When staff see their role as conserving specimens at all cost and researchers need to extract knowledge, even using destructive sampling, I believe that, with few exceptions, science should win. This is why some depth of duplication in collections is important, as well as remembering that the scientific value in specimens is measured in knowledge. This may sound cavalier, but it is not. Systematists, appreciative of both the information content of specimens and their fragility, approach their study with great reverence and all due caution.

The safest specimens are those that are locked away in cabinets, never handled or studied. They may be physically secure, but to what end? There are specimens with special significance deserving extra sensitivity, including types and those collected or studied by scientific greats like Banks, Owen, Darwin, Wallace, Cuvier, Buffon and others. But these are the minority. Only with constant taxonomic research is the information associated with specimens reliable and up to date. And only by studying specimens is the information content of collections tested and expanded, adding value. Natural history collections are, or at least should be, research resources that are responsibly used on a daily basis, not sterile archives infrequently consulted as historical artifacts. As we confront the biodiversity crisis, the growth of natural history collections is of paramount importance. Not haphazard collecting by inexperienced field hands, but targeted collecting directed by taxon experts. With so much extinction, knowledge of species will increasingly depend on museum collections. Nothing else we do will be of greater or longer enduring importance. Far from relics of the past, collections are keys to the future.

Systematics Ought to Conform to the Precepts of Other Sciences

Perhaps it is because so many biologists depend on identifications and scientific names to do their work that they feel justified judging systematics through the lens of their own discipline. The "New Systematics," born at the height of popularity of population genetics, tried to force taxonomy into the mold of a genetics world view. This was especially harmful to taxonomy. It confused the goals and methods of studies of populations with those at and above the species level, a confusion that persists to this day. The two complement one another but are different. Never fully recovered from damage done by such "population thinking," taxonomy now finds itself under the thumb of molecular genetics. It is critical to reassert the identity of systematics because, as Saunders said, "Taxonomy has never been more vital and less understood than it is today."

Biologists nurse an old inferiority complex. Longing for the precision of the physical sciences, many embrace DNA with the hope that reductionism will make biology look more like chemistry. In so doing, they hope of sharing in the prestige attached to physical laws and mathematical language. Taxonomy is better served by embracing the rich tapestry of evidence relevant to it and the unpredictable, contingent and fascinating nature of evolutionary history. Hypotheses in systematics today require no apology. Although operating in an environment different from that of physics, chemistry or genetics, ideas in systematics are as rigorously scientific as any. Reductionism is appropriate in some circumstances, but emergent properties of complexity in morphology make an increasing focus on the molecular level counterproductive to understanding species diversity and history.

Theodosius Dobzhansky's quote "Nothing in biology makes sense except in the light of evolution" is often cited as a retort to Creationism. With interesting implications for the current friction between descriptive systematics and molecular genetics, historian of science Mary Winsor has pointed out that Dobzhansky was actually concerned that organismal biologists were being disrespected. Cause for concern remains as we approach the study of biodiversity precisely backward. We need to reconstruct the sequence of character transformations, as they happened, then consider processes when and where inferences are justified. Instead, we continue to seek nonexistent laws of evolution with which we can understand evolution and avoid laborious scholarship. It will never happen. History has been shaped by too many forces, acting in unpredictable ways and combinations, and not infrequently upended by random events, to ever make it predictable. This is why analyses of DNA that make assumptions about evolutionary processes repeat mistakes made by evolutionary taxonomists decades ago.

Systematic biology will never be distilled to law-like rules. And reductionism will never replace deep knowledge of the emergent properties of complex morphological structures. We can either accept this and get on with the exploration of the diversity and history of life or give up. History is not reducible to mathematical formulae; it is too often shaped by chance events, and the species it has produced are individual kinds that could not have been predicted from their precursors. Attempts to disguise taxonomy as genetics will ultimately fool no one and diminish the excellence systematics is capable of achieving on its own. Let's stop trying to make the tail wag the dog by returning systematics to its own mission, being a good partner to genetics and the environmental sciences and accepting the challenge of discovering what biodiversity truly looks like.

Further Reading

Bell, N. L. (2002) A computerized identification key for 30 genera of plant parasitic nematodes. *New Zealand Plant Protection*, 55: 287–290.

Bremer, K., Bremer, B., Karis, P., and Källersjö, M. (1990) Time for change in taxonomy. *Nature*, 343: 202.

Darwin, C. (1859) *On the Origin of Species*, Murray, London, 502 pp.

DeSalle, R. and Goldstein, P. (2019) Review and interpretation of trends in DNA barcoding. *Frontiers in Ecology and Evolution*, 7: 1–11. doi:10.3389/fevo.2019.003

Emerson, R. W. (1950) Self-reliance. In *The Complete Essays and Other Writings of Ralph Waldo Emerson* (ed. B. Atkinson), Random House, New York, pp. 145–169.

Hennig, W. (1966) *Phylogenetic Systematics*, University of Illinois Press, Urbana, 263 pp.

Jha, A. (2005) Daisy has all the digital answers to life on Earth. *The Guardian*, 18 August.

Luc, M., Doucet, M. E., Fortuner, R., et al. (2010) Usefulness of morphological data for the study of nematode biodiversity. *Nematology*, 12: 495–504.

MacLeod, N., ed. (2008) *Automated Taxon Identification in Systematics*, CRC Press, Boca Raton, 368 pp.

Nelson, G. (2004) Cladistics: Its arrested development. In *Milestones in Systematics* (eds. D. M. Williams and P. L. Forey), CRC Press, Boca Raton, pp. 127–147.

Nixon, K. C. and Wheeler, Q. D. (1992) Extinction and the origin of species. In *Extinction and Phylogeny* (eds. M. J. Novacek and Q. D. Wheeler), Columbia University Press, New York, pp. 119–143.

Popper, K. (1959) *The Logic of Scientific Discovery*, Hutchinson & Company, London, 513 pp.

Raposo, M. A., Stopiglia, R., Brito, G. R. R., et al. (2017) What really hampers taxonomy and conservation? A riposte to Garnett and Christidis (2017). *Zootaxa*, 4317: 179–184.

Russell, K. N., Do, M. T., Huff, J. C., and Platnick, N. I. (2007) Introducing SPIDA-Web: Wavelets, neural networks and internet accessibility in an image-based automated identification system. In *Automated Taxon Identification in Systematics* (ed. N. MacLeod), CRC Press, Boca Raton, pp. 131–152.

Saunders, T. E. (2020) Taxonomy at a crossroads: Communicating value, building capability, and seizing opportunities for the future. *Megataxa*, 1: 63–66.

Wheeler, Q. (2004) Taxonomic triage and the poverty of phylogeny. *Philosophical Transactions of the Royal Society of London, B*, 359: 571–583.

Wheeler, Q. (2005) Losing the plot: DNA "barcodes" and taxonomy. *Cladistics*, 21: 405–407.

Wheeler, Q. (2010) What can we learn from 20th century concepts of species? Lessons for a unified theory of species. In *Für ein Philosophie der Biologie* (eds. I. Jahn and A. Wessel), Kleine Verlag, Munich, pp. 43–60.

Wheeler, Q., Bourgoin, T., Coddington, J., et al. (2012) Nomenclatural benchmarking: The roles of digital typification and telemicroscopy. *Zookeys*, 209: 193–202. doi:/10.3897/zookeys.209.3486

Wheeler, Q. and Meier, R. eds. (2000) *Species Concepts and Phylogenetic Theory: A Debate*, Columbia University Press, New York, 230 pp.

Wheeler, Q., Raven, P. H., and Wilson, E. O. (2004) Taxonomy: Impediment or expedient? *Science*, 303: 285.

Wheeler, Q. and Valdecasas, A. (2007) Taxonomy: Myths and misconceptions. *Anales del Jardin Botánico de Madrid*, 64: 237–224.Winsor, M. (2021) "I would sooner die than give up": Huxley and Darwin's deep disagreement. *History and Philosophy of the Life Sciences*, 43: 53. doi:10.1007/s40656-021-00409-3

9 The Species-Scape

What does biodiversity look like? This may sound like a silly question. After all, we see biodiversity most every day, even if only in the form of a city park or back garden. Even exotic habitats we may never experience are brought into our living rooms by television storytellers like David Attenborough. It turns out that there is no single answer to the question. How we respond depends on our perspective.

To an ecologist, biodiversity may be seen as the roles, interactions and abundance of species in a habitat. To a geneticist, the frequencies of traits in a population. To a landscape painter, a panoramic eyeful of nature. To a preacher, the grandeur of the Creation. To a farmer, a cultivated landscape through the seasons, occasionally raided by unwelcome weeds and pests. As a taxonomist, I have two complementary visualizations, one contemporary and the other historical. First is the species-scape. An imaginary landscape on which the sizes of organisms are proportionate to the numbers of species in the groups they represent. It is a snapshot of the diversity of species as we know it. And second, a cladogram or tree-like branching diagram—often, if imprecisely, called a phylogeny—that depicts the sequence of speciation events and associated character transformations. The information in cladograms is verbally expressed in Linnaean classifications and used to determine which organisms are depicted in the species-scape.

Either way, species-scape or cladogram, our current picture of biodiversity is incomplete and inaccurate. We do not know how many species exist within any major taxon and the vast majority of species are yet to be discovered. The living world we think we know is an illusion based on incomplete data and preconceptions. Only a comprehensive inventory and phylogenetic classification can bring the actual species-scape into focus.

Cladograms show relative relationships among species, not absolute ones: for example, two species, A and B, may be said to be more closely related to one another than either is to a third, C. Even if additional species are discovered to be nested between them, these relative relationships are unchanged. Nodes in cladograms represent hypothetical ancestors. Actual ancestors cannot be specified because no dispositive evidence exists by which to recognize them. Evolutionary novelties that might otherwise differentiate an ancestor are shared also by its descendants. While frustrating, cladograms depicting relative relationships are as

DOI: 10.4324/9781003389071-10

Figure 9.1 The Species-Scape. An imaginary landscape on which the size of each organism is proportionate to the number of species in the taxon it represents. Illustration: Frances L. Fawcett.

close as science can get to reconstructing the actual phylogeny that, of course, has real species (ancestors) at nodes and absolute closest relatives. Recognizing this limitation, and that "phylogeny" and "phylogenetic tree" have more precise meanings, they are nonetheless commonly used to refer to cladograms.

After centuries of exploration, we continue to make discoveries that revise our understanding of the diversity and history of life. In 1749 Linnaeus, the father of modern taxonomy, estimated the number of species:

> If we estimate the plants to approximately 10,000, the worms to 2,000, the insects to 10,000, the amphibians to 300, the fishes to 2,000 and the tetrapods to 200, then there are 26,500 species of living beings in the world.

At least two orders of magnitude too low, his guess nonetheless inspired generations of intrepid explorers to fan out across the globe in search of species. To date, more than 25,000 orchids (family Orchidaceae) and more than 90,000 weevils and relatives (superfamily Curculionoidea) have been named … and counting. Like Linnaeus, we do not yet have sufficient knowledge with which to confidently say how many species there are.

Even for well-known taxa in Europe, the most thoroughly explored continent, species continue to be discovered. Indeed, something on the order of 600 species are added to the continent's catalogue each year. All that we can say with confidence is that about 2 million species have been named to date and that the total is very much higher. Even in its neglected state, the taxonomic community names about 18,000 new species each year. Species, of course, take thousands to millions of years to evolve, so "new" really means newly known to science. There are also isolated populations that represent potentially new species in the making. While interesting to population biology, until they are fully fledged, they don't count. Some species not yet recognized by scientists are known to indigenous peoples who may have common names and uses for them. Other species have never knowingly been seen by a human being. A conservative estimate of 10 million species of plants and animals is reasonable, recognizing that the number could be much higher and that a comparable, perhaps greater, number of microbial species exist.

When we talk about numbers, insects are a big part of the conversation. At present, they account for half of all species. And, so far, this proportion is holding with roughly half of new species named each year being insects or related arthropods. Until the early 1980s, it was commonly thought that there might be a few million species. Terry Erwin, an entomologist at the Smithsonian, blew the lid off the subject when he announced his estimate of 30 million species of insects alone! If you have seen the movie *Arachnophobia*, you may remember the opening scene. A canister of pesticide is shot high into a tropical rainforest canopy, precipitating thousands of insects to rain down on collecting tarps outstretched below. That technique was perfected by Erwin who was thus able to sample a previously inaccessible, virtually unknown treetop fauna. Shocked by the number of new species in his samples, and by the high percentage of insect species unique to each kind of host tree, Erwin realized that the world has far more species than previously imagined. His calculation was as simple as it was startling. Take the average number of insect species associated with each kind of host tree, multiply by the number of tree species, and *voila*! His 30 million species estimate for insects has since been revised downward by most entomologists, but his work forever changed our perception of species diversity. My personal preference is to avoid overstating the scale of an inventory, then be pleasantly surprised if the numbers prove greater. Grimaldi and Engel, in their masterpiece on the evolution of insects, suggest that the 1 million named insects is perhaps 25% of the total. If this smaller number proves correct, 4 million is still a lot of insects.

Erwin's numbers unleashed a controversy that continues. Some say his estimate was unrealistically high, preferring a total in the 8–10 million range. Others propose totals as high as 100 million species, or more. A small industry of species number estimation has grown up, built on unverified assumptions, fragmental data and fanciful extrapolations. Some begin with the historical rate of species discovery or look for an asymptotic leveling off of new species in ecological samples. Others turn to opinions of seasoned experts who have devoted a lifetime to a particular group or to trends in monographs. In no case can we be confident whether assumptions or statistical gymnastics are realistic. We simply know too little. I personally suggest that we arrive at the number of species the old-fashioned way—by counting them.

It bothers me that the hoopla over species numbers has focused attention on the wrong question. Historians of science have found that asking the right questions matters more than finding correct answers. As Ed Wilson explained in *Consilience*, when you ask a trivial question, you get a trivial answer. But when you ask the right question, even if it cannot be answered exactly, you are led to major discoveries.

The number of species is the very definition of trivia. The right questions turn out to be the ones that taxonomists have been asking for centuries, the most basic of which is not "*How many* species are there?" but "*What* species are there?"

This question invites us to learn about species themselves, not simply count them. It elevates species from statistics to individual kinds, each worthy of close study. To learn *what* species exist, we must compare them and determine which characters make each unique. We must assess variation, geographic distribution and natural history traits as we go. An earnest pursuit of an answer results in the accumulation of a wealth of knowledge about species, characters and relationships. Such detailed knowledge should be our goal, not a number.

Were the important question really how many species exist, something like metagenomics could be a logical way to go, sampling disembodied DNA from the environment, detecting and counting species without ever seeing them or learning anything about them. Even given such a dumbed-down goal, DNA represents a grievous compromise. There is no formula by which genetic variation can be reliably translated into numbers of species. Given data on the average range of variation within related species, we can make an educated guess. But it is only a guess. Many closely related species are separated by genetic distances on the order of 2%. But there are exceptions, including cases in which greater variation exists within one species than between it and related species. Of the 15,741 gene families found in the common intestinal bacterium *Escherichia coli*, only 6% are present in every genome within the species. Species do not conform to some simple equation relating DNA diversity to numbers of species. That, of course, is why we call it biological *diversity*. Taxonomy treats species as carefully constructed hypotheses and repeatedly tests them. Molecular studies sometimes use the average genetic distance between sequences of known species as a yardstick for quick-and-dirty species "discovery." Any population differing by this average, or more, is proclaimed to be a species. Even though this often works to detect species, it is nonetheless a self-fulfilling claim. No matter how many times the distance is measured, species individuated by this arbitrary rule are not testable hypotheses.

As species are explored, knowledge of the spatial scale of the biosphere is changing, too. We used to picture the biosphere as a thin film of life stretched over the globe between a lifeless, rocky lithosphere below and uninhabited atmosphere above. Discoveries in recent decades have shattered this simplistic view of the biosphere. No one suspected chemical-energy-based ecological communities around deep-sea vents, springtails and round worms living thousands of feet below the surface in the world's deepest caves and mines, aquatic insects thriving in subterranean aquifers, fungus spores circulating 10,000 feet high in the air, diatoms living in hot springs or the biodiversity present in lakes isolated for millennia under the Antarctic ice cap. Add it all up, and we have a picture of the biosphere that is very different, much thicker and far more complex. We now appreciate that life penetrates deep into the oceanic and freshwater hydrosphere, high into the atmosphere and thousands of feet into the lithosphere. Rather than a thin veil, the biosphere turns out to be miles thick from top to bottom!

Because of the short span of a human life, our perception is that ecosystems are stable. Conservation rhetoric makes it sound as if we can stop change, maintaining the ecological assemblages of species that we are accustomed to in perpetuity. Like Sisyphus, this sets us up for an impossible task. The idea of permanent ecosystems is at odds with geologic history and realistic long-term expectations. Each generation accepts the *status quo* they are born into as an ecological baseline against which change is measured. Modest losses of species and degradation of ecosystems are not always perceptible over the few decades of a human life. Continue losses over a number of human generations and accumulated change can be huge. We are like the proverbial frog in a pot of water being brought slowly to a boil. It is among the tasks of science, and a compelling reason to complete an inventory of species, to create an ecosystem and biodiversity "memory," so that hard data exists to remind us what is happening over centuries and millennia. This does not mean that we can halt environmental change. But we can build a more complete picture of biodiversity in terms of current species and ecosystem composition. With such a baseline we can monitor change and better forecast, protect and adapt.

The forests I explored in my youth appeared to me to have been there forever. Parts of Ohio, where I grew up, were under an ice sheet a mile thick not much more than 12,000 years ago. My perception of the permanency of those ecosystems was false. Over geologic periods of time species are evolving in place, going extinct, immigrating and emigrating. Far from stable, ecosystems observed over long periods of time are kaleidoscopic assemblages of species being self-organized and reshuffled, over and over again. Similarly, genetic variation within single species is in constant flux with mutations arising, spreading and being eliminated by natural selection. It is the job of systematics to tease apart characters that are informative at and above the species level from a fog of genetic variability within species that, while interesting for other reasons, tells us nothing about the individuality of, or relationships among, species. This is why population genetics and taxonomy are distinct sciences. The sources of natural selection are varied and unpredictable, acting singly or in concert within a lineage. The thing that does not change is history—the irreversible pattern of relationships among species, evidenced by fixed, shared-derived characters.

When I use the word biodiversity I do so in a special sense referring to the similarities and differences among species and clades. Others use the word to refer to any and all sources of variation within or among organisms at any level, from molecules to ecosystems. In systematics, and in this book, we are concerned specifically with the diversity of life *at and above* the species level, what is sometimes called the macro-evolutionary or phylogenetic context. Geneticists are concerned with processes from replicating DNA to shifts in gene frequency in populations. Most of that is, for the taxonomist, noise. Taxonomists must be vigilant so as not to confuse variable traits with informative characters. The interest of systematists really begins where that of the population biologist ends, the point at which species are fully diverged and characters completely transformed. This is why systematics is historical, looking backward to explain the pattern of distribution of characters; why it is non-experimental and cannot unequivocally make assumptions about evolutionary processes. Those are best left to experimental fields where such phenomena may be observed in action. And a reason to be suspicious of analyses of data that make evolutionary process assumptions.

Returning to the question of what biodiversity looks like, odds are your first thought is of a landscape, an ecosystem viewed from a distance. Perhaps the irregular skyline of a forest canopy in the Amazon, big game at a watering hole on an African savannah or a collection of bushes and birds in your backyard. For most of us, the image in our head is as much fantasy as reality: a vague, green vegetative backdrop animated by a few mammals scurrying across the ground and birds flitting in the air. Viewed through different lenses, the living landscape looks very different.

One concern of ecologists is the abundance of organisms in a habitat. In a classic study of a pasture near Cambridge, England, George Salt and his colleagues meticulously counted the arthropods present in samples of the top 12 inches of soil. They found more than 263,000 arthropods per square meter, or 1,400,000,000 per acre! Included in the long list of arthropod residents were 248 million springtails, nearly 18 million beetles and about 666 million mites. Having dispelled the superficial image of a herd of cattle grazing on grass, you may never look at a pasture the same way again. I doubt that Salt ever did.

Even this seemingly extreme view of biodiversity, measured in numbers of individuals, fails to fully flesh out the picture. Many organisms have what ecologists call an "r" strategy of reproduction. Rather than investing in parental care and giving offspring the best possible chance to survive to maturity and reproduce (the alternative "K" strategy), they generate massive numbers in a game of chance. An example is the giant puffball, *Calvatia gigantea*, a single individual of which can release as many as 7 trillion spores into the world! If every spore of every puffball in one season survived to produce a fruiting body, each about half the size of a basketball, we would soon be up to our necks in fungi.

Species are so diverse and numerous that it is challenging to picture them. Years ago, catching up on reading during a flight home to Ithaca, a simple line drawing by Katherine Brown-Wing caught my eye. There was a giant ant (*Gnamptogenys pleurodon*) towering over a jaguar (*Panthera onca*)—a surprising image, begging explanation. As I read the associated text by E. O. Wilson, I realized that this was an elegant graphic representation of the biomass of inhabitants of a Brazilian tropical forest. Wilson's point was that, in a comparable area of forest, the biomass (i.e., dry weight) of social insects is about four times that of all vertebrate animals combined, including mammals, birds, reptiles and amphibians.

As a taxonomist, my first thought was to wonder what an analogous image would look like that depicted numbers of species grouped by their relationships. I pulled together the figures and, with scientific illustrator Frances Fawcett, produced what we called a "species-scape." This fanciful landscape was a visual challenge to a world I thought I knew.

Our first version was a foldout pen-and-ink drawing published in an invited article on insect diversity for the Entomological Society of America. We were unprepared for the surge of interest in the image following its appearance in the

Tuesday science section of the *New York Times*. It was subsequently reproduced scores of times, won a "Best of What's New" medal from *Popular Science* and eventually came full circle as an image in E. O. Wilson's *The Diversity of Life*. Frances would later produce both a computer-generated version of the species-scape and a beautiful painting that hangs in my study.

The species-scape reveals that, at least in species numbers, and contradicting our preconceptions, we do not live on a planet dominated by trees, mammals and birds, or even human beings. Viewed through the lens of species diversity, our world is dominated instead by microbes, fungi and animals with more than four legs … or no legs at all. If anything, given our present knowledge we live on the planet of the insects. In any event, the species-scape is wholly foreign to us, something from science fiction, a landscape as weird and unexpected as what astronauts might find traveling to a distant planet.

The real species-scape is so fantastic that I'm reminded of Groucho Marx who asked "Who are you going to believe, me or your own eyes?" In this case, believe me. While mammals loom large in our imaginations, because of their large size, impressive diversity and relationship to us, they account for only about six and half thousand species. Birds, even under the phylogenetic species concept that is estimated to double their number, total only about 18,000. Compared to hundreds of thousands of flowering plant species, or more than 1 million insects, they are minor players in the species number game. In all mega-diverse groups, taxonomists are greatly outnumbered by unknown species. The pace of species discovery is not limited by the supply of undescribed species, only the number of experts and amount of effort and resources devoted to discovering them. Sadly, as we ignore taxonomy and the core mission of natural history museums, a generation of species experts is dying off without passing their specialized knowledge on to a new generation of species explorers. A fraction of what they know is recorded in publications and curated collections, of course, but a wealth of additional knowledge has traditionally been shared in apprenticeship-like fashion, and over a few pints at professional meetings. This knowledge can be recreated through the school of hard knocks, but this wastes valuable time that could be spent moving forward.

As strange as the species-scape is, it too is illusory. Current proportions of organisms are far from accurate. As massive as the beetle is, it represents only a little more than 1 million kinds of insects that have been named. Not the 3 million or more awaiting discovery. Other groups are even farther out of proportion, such as mites, round worms and fungi, some of which may overshadow insects when fully known. Only a complete inventory of species can bring the actual species-scape into focus and allow us to see the world of species diversity as it really is, for the first time. Given the rate at which species are being driven to extinction, we owe it to ourselves to document the species-scape while we can. It is being transformed rapidly with tens of thousands of species erased from the image each year. To understand the species-scape, and how it is being redrawn by extinction, we urgently need to address the great questions asked by systematics.

Further Reading

Barrowclough, G. F., Cracraft, J., Klicka, J., and Zink, R. M. (2016) How many kinds of birds are there and why does it matter? *PLoS ONE*, 11: e0166307. doi:10.1371/journal.pone.0166307

Burgin, C. J., Colella, J. P., Kahn, P. L., and Upham, N. S. (2018) How many species of mammals are there? *Journal of Mammalogy*, 99: 1–14.

Erwin, T. L. (1982) Tropical forests: Their richness in Coleoptera and other arthropod species. *Coleopterists Bulletin*, 36: 74–75.

Essl, F., Rabitsch, W., Dullinger, S., et al. (2013) How well do we know species richness in a well-known continent? Temporal patterns of endemic and widespread species descriptions in the European fauna. *Global Ecology and Biogeography*, 22: 29–39.

Grimaldi, D. and Engel, M. S. (2005) *Evolution of the Insects*, Cambridge University Press, Cambridge, 755 pp.

Hestmark, G. (2000) Oeconomia naturae L. *Nature*, 405: 19.

Hodkinson, T, R. and Parnell, J. A. N., eds. (2006) *Reconstructing the Tree of Life: Taxonomy and Systematics of Species Rich Taxa*, CRC Press, Boca Raton, 368 pp.

Larsen, B. B., Miller, E. C., Rhodes, M. K., and Wiens, J. J. (2017) Inordinate fondness multiplied and redistributed: the number of species on earth and the new pie of life. *The Quarterly Review of Biology*, 92: 229–265.

Linnaeo, C. (1749) *Oeconomia naturae*, Upsala, 56 pp. linnean-online.org/120097/

Lukjancenko, O., Wassenaar, T. M., and Ussery, D. W. (2010) Comparison of 6 sequenced *Escherichia coli* genomes. *Microbial Ecology*, 60: 708–720.

Raven, P. H., Gereau, R. E., Phillipson, P. B., et al. (2020) The distribution of biodiversity richness in the tropics. *Science Advances*, 6. doi:10.1126/sciadv.abc6228

Salt, G., Hollick, F. S. J., Raw, F., and Brian, M. V. (1948) The arthropod population of pasture soil. *Journal of Animal Ecology*, 17: 139–150.

Stork, N. E., McBroom, J., Gely, C., and Hamilton, A. J. (2015) New approaches narrow global species estimates for beetles, insects, and terrestrial arthropods. *Proceedings of the National Academy of Sciences*, 112: 7519–7523.

Wheeler, Q. D. (1990) Insect diversity and cladistic constraints. *Annals of the Entomological Society of America*, 83: 1031–1047.

Wilson, E. O. (1990) *Success and Dominance in Ecosystems: The Case of the Social Insects*. Excellence in Ecology, 2., Ecology Institute, Oldendorf/Luhe, 104 pp.

Wilson, E. O. (1992) *The Diversity of Life*, Belknap/Harvard University Press, Cambridge, 424 pp.

Wilson, E. O. (1999) *Consilience: The Unity of Knowledge*, Vintage Books, New York, 367 pp.

10 The Illusion of Knowledge

Librarian of Congress Daniel Boorstin said that "The greatest obstacle to discovery is not ignorance, it is the illusion of knowledge." That illusion includes the current flood of data. NASA expects to have more than 250 petabytes of environmental data by 2025. A petabyte is 1 million gigabytes! And, the National Center for Biotechnology Information announced in October, 2022, that GenBank release 252.0 had reached the milestone of 20 trillion bases from 3 billion records. This fog of data, in spite of its usefulness, is blinding us to things we don't know. We mistake quantity of data for quality of knowledge, but more is not always better. As a molecular-based approach to taxonomy gains in popularity, it creates the illusion of more knowledge of species than exists.

A number of variations on molecular-based taxonomy have been proposed and, thankfully, so far rejected in favor of approaches that integrate data sources. But the lure of a DNA-based system is strong and efforts persist. One recent "minimalist" protocol proposed by Meierotto et al. consists of an image of a specimen, a *COI* barcode diagnosis and the designation of a holotype. This is like curing a cold by executing the patient. It achieves the goal of stopping the virus but misses the point of practicing medicine. The goal of taxonomy is to learn the results and history of evolution at the granularity of species, characters and clades, not simply to make slap-dash identifications. Skipping over detailed descriptions, we miss that which is most interesting and informative. DNA barcode advocates have succeeded in presenting to systematics a Trojan horse, giving lip service to descriptive taxonomy while failing to share research funds or advocate in good faith for its support.

If taxonomy is not done right the first time, it is doubtful that "maximalist" descriptions will ever exist for most species. Minimal descriptions may address ecologists' immediate need for identifications, but they ignore our last best opportunity to document earth's species. There is no excuse for a minimalist approach when in-depth descriptions could be done efficiently to high standards. We simply need to support taxonomists to do so. It is odd that the same people who claim there are too many species to describe propose to barcode them all. The additional work involved in making detailed descriptions, hypothesis-based species and comprehensive collections is insignificant compared to the knowledge lost by failing

DOI: 10.4324/9781003389071-11

to do so. Not only can we describe every species to a high standard, we can do so in a matter of decades.

Humankind has done pretty well for itself knowing a small fraction of earth's species, so why make the effort to learn about the others now? In earlier times, the same could have been said of just about any advance. Learning that the earth was round was a boon for navigation, but life could have gone on under the misconception that the world is flat. And we muddled along for most of the time that humans existed without understanding properties of chemical elements. But, considering the enormous benefits of understanding them, why would we choose to remain ignorant? Yet, when it comes to species, the greatest unexplored frontier on our planet, that is exactly what is happening. We are voluntarily settling for measures of DNA diversity when deep knowledge of species is ours for the taking.

We could write off exploring species and get along, at least for a while. We could feel noble for our sentiments toward biodiversity, never knowing what species had been saved or lost. Ignorance of evolutionary history limits our understanding of ourselves and our world but is no threat to our survival. But, given the option to live sustainably, effectively conserve species and understand the origins of biodiversity, why would we choose not to?

Living in North America or Europe, it is easy to assume that we know more about species than we do. Nearly all plants and animals we encounter in our daily lives have names and can be identified, more or less readily. We associate discoveries of new species with far away islands and tropical rainforests. But each of us has, possibly today, brushed by species unknown to science. There is no location in the world where every single species is known, at least not when microbes are included. Countries with relatively few species, the best collections and the longest traditions of taxonomy are comparatively better known, of course, and yet they, too, have species regularly discovered.

Field biologists living in or visiting the most species-rich regions of the globe realize quickly just how little we know. The majority of species in so-called biodiversity hotspots—places like the Amazon and Congo basins, Indonesia and Myanmar—have no scientific names and cannot be identified. We know nothing of their morphology, ecology or natural history aside from what we can infer from related species. We are well advised to recall that discoveries of individual species have changed the course of history and will again. Think of the impact on the human condition of discovering maize (*Zea mays*), mosquito vectors of Malaria (e.g., *Anopheles*) or the olive tree (*Olea europaea*) the oil of which enriched nutrition and pierced the night. These are but three among tens of thousands of examples. Fractional knowledge of species contributes trillions of dollars to the world's economy. Undiscovered species represent a vast untapped resource with the potential to dramatically improve and enrich our lives. Yet, we have done little to formalize a campaign to discover them. Instead of commonsense investments in systematics, we focus on technological alternatives that yield minimal information. This is a failure of leadership from institutions that value profits and popularity more than the creation of knowledge and doing the right thing.

Support for systematics is an investment in possibilities. Pure science recharges the well of options for problem-solving. Focusing only on today's problems limits us to today's solutions. Because of the diversity of life, I cannot imagine any area of research more capable of producing new and unexpected opportunities than taxonomy. But the option to explore, document and describe species is disappearing as we foolishly ignore the needs of systematics.

Joel Cracraft recognized seven great questions asked by systematic biology. The pursuit of these questions remains essential to our understanding of biodiversity, the biosphere and evolutionary history. Rampant species extinction means that the time to seek answers is growing short. I have slightly rephrased some of Cracraft's questions and added an eighth. If we play our cards right, taxonomy will remain a reliable fountainhead of ideas with which to create a sustainable future for people and the natural world.

What Is a Species?

Although there is not yet consensus among biologists, I believe that we have arrived at a general species concept that is theoretically sound and broadly applicable. After centuries of debate, we are at last approaching a unified species concept for all living things. This concept is surprisingly similar to the first formal definition of species by John Ray in 1686:

> no surer criterion for determining species has occurred to me than *the distinguishing features that perpetuate themselves in propagation from seed*. Thus, no matter what variations occur in the individuals or the species, if they spring from the seed of one and the same plant, they are accidental variations and not such as to distinguish a species .... Animals likewise that differ specifically preserve their distinct species permanently; one species never springs from the seed of another nor vice versa
>
> (John Ray, 1686, italics mine)

> We define species as the smallest aggregation of (sexual) populations or (asexual) lineages *diagnosable by a unique combination of character states*.
>
> (Wheeler and Platnick, 2000, italics mine)

The road to a species concept applicable to all organisms was complicated by assumptions that were, in retrospect, unnecessary. Some concepts were based on a favorite evolutionary process, while others were tailored for a particular group of organisms. Not surprisingly, concepts optimized for one process or taxon proved less well suited, or entirely inapplicable, to others.

The secret to a unified concept of species is deceptively simple: base it on what all species share in common, avoiding an emphasis on things in which they may differ. Life is so diverse that about the only thing shared by every species is evolutionary history. Because history was shaped by numerous processes and events, it is the pattern of attributes in and among species that is key. Every species, regardless of the processes that played a role in its origin, belongs to one history. The goal is to recognize species as the outcomes of that history, something accomplished by studying patterns of character distributions. Unique combinations of characters are indicia of species. As patterns, they are compatible with all conceivable evolutionary processes and limited by none. Once recognized, species become the elements—least inclusive, indivisible units—of phylogenetic analysis.

The only realistic alternative to a historical, pattern-based concept is pluralism, using different concepts for different groups of organisms. While this works well for some purposes, it robs us of interesting insights into evolutionary history, biodiversity and the organization of the biosphere. Unless species are comparable units across all taxa, we cannot ask whether there are more kinds of beetles or bromeliads, whether one evolutionary process produces greater or lesser numbers of species than another or whether there are more species in a hectare of Amazonian rainforest or Arctic tundra. Biologists ask these questions, usually ignoring the fact that different concepts in different taxa may skew the answers. Birds are a good case in point. Because the biological species concept, based on assumptions about interbreeding, has been preferred in ornithology, the number of species has been undercounted. Barrowclough and colleagues, applying the phylogenetic species concept, estimated that the actual number of bird species may be about twice the current count; in round numbers, closer to 18,000 than 9,000. For avid birders with few conquests remaining for their life lists, this should come as good news. More importantly, it aligns birds with other taxa based on the phylogenetic species concept so that broad questions about biodiversity may be asked.

In spite of the protracted search for a general species concept, stymied on occasion by complex examples, a large number of species are plainly obvious to anyone paying attention. Jared Diamond found that native people in New Guinea had common names for all but 1 of the 136 bird species recognized by professional ornithologists—no formal theoretical concept or graduate degree required, just a keen eye.

For nearly a century, the "biological" species concept dominated zoology. Based on the ability to interbreed, it has been religiously taught to generations of high school and college biology students. Surprisingly, few zoologists professing to use this concept have ever put it to the test, and botanists smartly rejected it because of frequent hybridization among plant species.

For me, the death knell for the biological species concept was rung by ichthyologist Donn Rosen. I vividly remember the day I read his paper and a bulb lit in my head. Rosen was studying species of swordtail fishes of the genus *Xiphophorus* in Guatemala. If you have ever kept a freshwater aquarium, you may have had some pet swordtails. Males, as their name suggests, have an impressively long, swordlike lower lobe of the caudal fin. In nature, they inhabit isolated streams in karst terrain of Central America where they never come into contact with one another. Rosen brought live specimens back to his lab at the American Museum of Natural History in New York City and undertook all the possible crosses.

What Rosen found was that some, but not all, fully differentiated species could successfully interbreed when artificially given the opportunity. This was true even

Figure 10.1 A Swordtail Fish. *Xiphophorus montezumae*, male, lateral view. Photo: Courtesy of the *Xiphophorus* Stock Center, Texas State University.

of species separated from one another by more than one intermediary branch in the phylogenetic tree. Rosen's conclusion was that the ability to interbreed is a primitive characteristic: all individuals in the original *Xiphophorus* species were capable of breeding with one another. What really matters is whether descendants have sufficiently diverged so that they can no longer interbreed. Stated another way, it is the loss of the ability to reproduce that is an evolutionary novelty, and therefore informative, not the ability to interbreed as the so-called biological concept dictates. We should not be surprised when closely related species, normally isolated in nature, are able to interbreed when artificially brought together.

Rudolf Meier and I invited scientists holding several popular species concepts to engage in a written debate. Each author, or pair of authors, was invited to write an introductory chapter explaining the tenets of their concept. Once submitted, manuscript changes were not allowed beyond grammatical corrections. A second set of chapters consisted of critiques of competing concepts. A third, and final, set of chapters then rebutted attacks. This debate format worked well. Unlike a literature review, readers were led through point/counter-point arguments by authors who genuinely believed in the concept they wrote about and left to draw their own conclusions. If I say so myself, Norman Platnick and I clearly won the debate. Inexplicably, the other authors continue to see things differently. For those interested in a deep dive into issues associated with species concepts, the book remains a useful introduction.

To summarize, any concept based on a specific process, or optimized for a single taxon, cannot function equally well for all organisms. Life is too diverse, processes too many. And there are compelling reasons to want a uniformly applicable concept, such as comparing rates of speciation or numbers of species in clades or ecosystems. A general species concept must allow for any and all evolutionary processes and taxon peculiarities. The only way to achieve a concept applicable to all life is, therefore, to focus on patterns, not processes. The goal of a general or unified species concept is to recognize kinds of organisms that are outcomes of evolutionary history, diagnosable by a unique combination of heritable characters and theorized to be consistently and irreversibly different from all other kinds. This must be tempered by recognizing that species are hypotheses, not facts. We must continually search for more characters and observe additional specimens to test species hypotheses and assure that they indeed reflect patterns in nature. It is impossible to know with certainty that species are fully and irreversibly differentiated, so the best science can do is to make testable claims.

Two swordtail species, isolated in different streams and consistently differing from one another today, if capable of interbreeding, might merge into a single species at some point in the future were their drainage systems to converge through some geological event. Given available evidence, Rosen's species represent the best hypotheses of the situation as it exists. New evidence, or changing realities on the ground, could require changes in hypotheses at a future date. This is the nature of science. No absolutes. No guarantees. Just ideas formulated so that they are consistent with available evidence and can be tested.

What Species Are There?

I have changed this question from "*How many* species are there?" to "*What* species are there?" As previously mentioned, asking what species exist invites us to discover and describe what makes each species unique and how its attributes compare to those of other species. The goal of systematics is to know and classify species, not just learn that they exist.

If we literally wanted to know only how many species there are then, assuming that certain challenges could be overcome, metagenomic environmental surveys might suffice. A few years ago, Craig Venter set sail with the goal of sampling sea water, isolating DNA and estimating how many species live in the oceans. News articles announced that Venter had discovered more species than anyone in history. I was incensed. Simply demonstrating lots of genetic variation has little to do with discovering species. Even if we had a reliable formula to translate genetic diversity into species numbers—and no such formula exists—I would hesitate to call his discoveries "species" in any meaningful sense. As a taxonomist, I have higher standards. I expect to know more about supposed species than genetic variation before claiming to have discovered anything. Minimally, species ought to be described and presented as explicitly testable hypotheses. In respect to numbers, take your pick. I prefer to be conservative and follow Chapman and others who, after consulting with experts, have thoughtfully arrived at 10 million species as a nice round number. The reality is that we simply do not know—yet.

What Is the Tree of Life?

We know that evolution happened. We know many of the processes involved. But we know astonishingly little about what actually took place and what resulted. We have only begun to learn the species that exist, much less the chronology of character transformations and speciation events responsible for the diversity of life.

In science, we typically learn that something took place then investigate how it happened. For evolution, we have approached things backward. In the time since Darwin, we have focused on how the mechanisms of evolution work, from the molecular basis of inheritance to forces of natural selection. But our knowledge of what actually happened is scant. This is like studying the social, economic and political structures of ancient Athens and Rome without knowing which civilization was dominant first. Such factors indeed shaped these societies, but their explanatory power means little in the absence of an accurate chronology.

We have made impressive progress working out relationships among major branches in the tree of life, but our knowledge of smaller limbs, twigs and leaves is nascent. Fascinating details of history remain to be unveiled before we can begin to truly understand phylogeny. This means studying relationships at every level from species to life as a whole. And it means doing so at the granularity of individual characters, not relying on indirect measures like percentages of genetic similarity.

A telling feature of the motives of many molecular studies is the publication of naked trees. They show branching relationships among species but have been stripped of all the interesting stuff: the attributes of species that the tree claims to explain. This visible lack of interest in the properties of species themselves should be disturbing to anyone curious about the diversity of life and its origins.

The importance of the pattern approach of systematics is made clear by the diverse factors shaping the course of evolution. Stephen Jay Gould beautifully illustrated the contingency of evolution in his discussion of the bizarre fauna of the Burgess shale formation in *Wonderful Life*. A perfectly well-adapted species can be wiped out by a chance meteor strike or some other catastrophe. Such random events cannot be anticipated, yet can exert profound influence over the future direction of evolution. This means that, regardless of how well we understand processes of mutation and selection, it remains necessary to reconstruct what actually took place independent of process assumptions.

Phylogeny, like human history, is a one-off. Even if we could rewind the clock, restarting evolution with the same actors, at the same time, under precisely the same conditions we would still get a different outcome. Those who like tidy, law-like predictability will find studying phylogenetic history frustrating. Try as we may, evolution cannot be explained by rigid rules. Molecular clocks sometimes skip a beat. And we must expect random occurrences of the unexpected. For me, this makes systematics more intellectually exciting than would be the case if evolution were constrained by entirely predictable laws.

Predictions about phylogeny are tested each time new evidence is introduced, whether a previously undetected character or simply a newly examined specimen. In order to refine and corroborate our understanding of history, it is imperative that we gather and preserve as much evidence as possible. This includes museum specimens of species soon to disappear in the wild.

What Is the History of Character Transformations?

Species and characters are two sides of the same coin. Phylogenetic history is understood in terms of hierarchic relationships among both characters and the species bearing them. This is why descriptive taxonomy is so important. It drills down to individual characters that are both evidence of species and common ancestry, and fascinating outcomes of evolution. Describing the attributes of species is a necessary first step if we wish to understand details of evolutionary history. With most species undescribed, and others inadequately described, much work remains.

Richard Dawkins insightfully said, "The essence of life is statistical improbability on a colossal scale." This really gets to the point. It is the unlikely, seemingly boundless diversity of life, strikingly evident in comparative morphology, that makes species exploration and the study of evolutionary history so intellectually rewarding. If we were not consistently rewarded with weird and unexpected characters as we describe species, we could be forgiven for pursuing molecules at the expense of morphology. But we are, and to ignore descriptions is to gloss over the greatest rewards of species exploration.

Where Are Species Distributed, Geographically and Ecologically?

Knowing geographic and ecological distributions of species is important for understanding biodiversity as well as managing natural resources, planning land use and establishing conservation goals. Species geography has contemporary and historical components. Ecologists assess biotic and abiotic factors limiting ranges today, while biogeographers compile evidence of geographic patterns, past and present. Understanding spatial patterns requires that we take into account ecological, topographical and climatological constraints on present day distributions as well as accounting for influences of history. Species occur where they do, in part, because their ancestors lived or migrated there.

Vicariance biogeography, a branch of historical biogeography, is interested in patterns repeated among unrelated taxa that may point to events affecting entire biota. The initial discovery of tectonics, that continents are riding on floating plates, was based in part on animal and plant distributions. It was curious that species in South America had their nearest living relatives in Africa or Australia rather than North America, which is much closer. As the Gondwana supercontinent broke apart beginning in the early Jurassic, the fragments of its biota account for this pattern repeated among diverse taxa.

How Have Species Distributions Changed Through Time?

Just as cladistic analyses tell us the sequence of species origins and character transformations, historical biogeography informs us regarding changes in distributions through time. Such retrospective geography is not only an important part of understanding the world as we find it, it is the basis for anticipating coming changes. An example is the use of fossil species to study past climate change, such as pioneering work on freshwater midges by Steve Brooks of The Natural History Museum in London. These tiny flies of the family Chironomidae have aquatic larvae whose sclerotized head capsules are preserved in lake bottom sediments. Brooks has calibrated temperature regimes for various midge species, tracking fluctuations in climate at late glacial and Holocene sites to the precision of 1°C. Such knowledge is invaluable as we seek to understand, monitor and predict climate change and its impacts.

How Can Classifications Be Predictive?

Phylogenetic classifications are not only an optimal system for information storage and retrieval, they are predictive. When a character is discovered in one species, the same or a similar character is more likely to be found in related species than at random among unrelated species. Discovering a new species of the spider family Araneidae, and knowing nothing about its habits, we can predict that it builds spiral, wheel-shaped "orb" webs, first with a framework of non-sticky silk, then with a final covering of silk and sticky droplets; has eight eyes; hairy or spiny legs; and lacks stridulatory organs. And we will almost certainly be correct. This ability to predict attributes of species we have not yet studied in detail is of immense value as we explore biodiversity, find our way around the biosphere and become more intentional in the pursuit of biomimicry. Attributes that are labile, such as behavior, may be more difficult to predict, yet phylogenetic classifications are a good starting point.

How Can Taxonomic Knowledge Be Leveraged for Biomimetic Solutions?

The biosphere is changing rapidly. Species extinction, exploitation of resources and climate change, among other assaults, are creating environmental problems faster than we are adapting to them. We need new, more efficient and less wasteful strategies to meet human needs and the ability to rapidly respond to environmental challenges as they arise. Our best hope is biomimicry, looking to nature for clues, models and inspiration to create ways to live more sustainably. Descriptive taxonomy can exploit a cumulation of ideas for biomimicry, but attributes of species must be organized in a searchable database and made accessible in terms understandable to engineers, designers, inventors, entrepreneurs and others outside the taxonomic community. A partnership with information scientists is needed to translate the technical vocabulary used by taxon experts into plain language. As cyberinfrastructure is designed to renew descriptive taxonomy, it should make provisions to mobilize knowledge for biomimicry.

How much do we need to know about species? Someday, after we have survived the biodiversity crisis, described and mapped life on our planet, conserved diverse species and become a sustainable society, scholars should study how it was that we could ask such a question. It may not be said out loud, but the resistance to morphology-based studies suggests that many scientists are content with bare minimum knowledge. We do not ask if we know enough about medicine, astronomy or physics, it being obvious that we are on a journey of learning that has no end. Considering our ignorance of species and evolutionary history, the paucity of options for adapting to environmental change and all that is at stake as species go extinct, it is shocking that so many scientists are satisfied to simply detect species with DNA while learning almost nothing about them. A serious pursuit of the eight great questions of systematics will lead us to profoundly important insights, including the evolution of morphological diversity.

Further Reading

Barrowclogh, G. F., Cracraft, J., Klicka, J., and Zink, R. M. (2016) How many kinds of birds are there and why does it matter. *PLoS ONE*, 11: doi:10.1371/journal.pone.0166307

Brooks, S. J. (2006) Fossil midges (Diptera: Chironomidae) as palaeoclimatic indicators for the Eurasian region. *Quaternary Science Reviews*, 25: 1894–1910.

Chapman, A. D. (2009) *Numbers of Living Species in Australia and the World*, Australian Government Department of the Environment, Water, Heritage and the Arts, Canberra, 80 pp.

Cracraft, J. (2002) The seven great questions of systematic biology: An essential foundation for conservation and the sustainable use of biodiversity. *Annals of the Missouri Botanical Garden*, 89: 127–144.

Darwin, C. (1859) *On the Origin of Species*, Murray, London, 502 pp.

Dawkins, R. (1986) *The Blind Watchmaker*, W. W. Norton, New York, 496 pp.

Diamond, J. M. (1966) Zoological classification system of a primitive people. *Science*, 151: 1102–1104.

Hennig, W. (1966) *Phylogenetic Systematics*, University of Illinois Press, Urbana, 263 pp.

Mayr, E. (1980) *The Growth of Biological Thought*. Harvard University Press, Cambridge, 974 pp.

Meierotto, S., Sharkey, M. J., Janzen, D. H., Hallwachs, W., et al. (2019) A revolutionary protocol to describe understudied hyperdiverse taxa and overcome the taxonomic impediment. *Deutsche Entomologishche Zeitschrift*, 66: 119–145.

Ray, J. (1686) *Historia Plantarum Generalis*, as translated by E. Silk in Mayr, E (1982), *The Growth of Biological Thought*, Harvard University Press, Cambridge, pp. 256–257.

Wheeler, Q. D. and Platnick, N. I. (2000) The phylogenetic species concept. In *Species Concepts and Phylogenetic Theory: A Debate* (eds. Q. D. Wheeler and R. Meir), Columbia University Press, New York, pp. 55–69.

Wheeler, Q. D. and Meier, R., eds. (2000) *Species Concepts and Phylogenetic Theory: A Debate*, Columbia University Press, New York, 230 pp.

11 Morphology without Apology

In an age of abundant, inexpensive molecular data why should we continue to study morphology? Able to identify species and reconstruct relationships with DNA, isn't morphology passé? I once heard the head of a molecular lab in a major natural history museum argue that we should no longer add specimens to collections, only tissues for DNA extraction. With appropriate instruction, and access to a properly equipped laboratory, it is possible to isolate and sequence DNA while knowing nothing about species themselves. There is no denying that morphology requires more knowledge, scholarship and labor. Or that glossaries of morphological terms are daunting to master. But morphology yields qualitatively valuable insights. Molecular data, even whole genomes, occupies a phylogenetic penumbra between the darkness of ignorance and evolutionary enlightenment. The latter comes only with knowledge of species and characters, understanding evidence, molecules included, in a broader biological context. As suggested in the words of Crowson, the present glorification of technology in no way trumps the fact that "The systematic importance of information is in no way dependent on its novelty, or on the sophistication or complexity of the techniques by which it was obtained."

A synthesis of evidence is best, but able to have only one source I would choose morphology. In most lineages, intricacies of morphology are among the most interesting outcomes of evolution. Include knowledge of the function of structures and it may be possible to infer habits from form. And morphology allows comparative studies of living and fossil species. There are examples of DNA recovered from extinct species, such as from the tooth of a mammoth encased in permafrost for 1.6 million years by van der Valk et al. But for most extinct species this is not a possibility. Fossils provide a glimpse of species and characters that no longer exist, but only a fraction of characters are preserved. Thus, in general, extant species yield far more evidence. All relevant data sources have value and even if we could get by with molecular data alone, why would we?

Since the time of Aristotle, making sense of the pattern of morphological similarities and differences among species has been a goal. To abandon or minimize morphology now is to fundamentally alter the core mission of systematics. Another argument in favor of morphology is common sense. When you visit a natural

DOI: 10.4324/9781003389071-12

history museum, botanical garden or zoo, when you rise at dawn to go birding or hike through a meadow to enjoy wildflowers, you do so to observe morphology. Such outings are most rewarding if you understand what you are looking at; when morphological details allow you to appreciate species diversity, and when a gallimaufry of morphology suddenly makes sense in the context of a phylogenetic classification.

Curiosity about body parts has spawned entire disciplines, from paleontology to genetics, embryology and evolutionary biology. Describing, comparing and understanding morphology will remain a central part of systematics for as long as the science is done for its own sake. One can only ignore or marginalize morphology if your motives are other than understanding species. Given centuries of progress, redefining taxonomy now, in the middle of a biodiversity crisis, is like rewriting the operator's manual for *Titanic* as she sinks. Efforts to make classifications "natural" were primarily driven by the desire to order morphological diversity in an evolutionary context. Nothing about this changed with the addition of molecular data, except that we have one more tool fit to purpose.

It is conceivable that we could, sampling tiny flakes, analyze pigments in paintings to create isotopic profiles by which we could tell them apart. Recording them as UPC barcodes, glued to the back of canvases, we could hang paintings face to the wall and anyone with a barcode reader could stroll through the Metropolitan and accurately identify paintings. But to what end? Is this why we visit art museums? Of course not. If such a silly approach were taken by art curators, we would either revolt or stop visiting museums. We want to see works of art for ourselves, to be emotionally touched by images, to marvel at brilliant compositions and be impressed by the imaginations and techniques of painters. Each canvas is a creation best admired in its full complexity, not merely identified. Push aside the technological whistles-and-bells, the lucrative research grants and the prestige of being perceived to be at the cutting edge of science, and this is more or less what is on offer from molecular systematics taken to an extreme. Going molecular, at the exclusion of morphology, misses the bigger picture. Masquerading taxonomy as genetics and offering quick identifications to ecologists may gain the respect of some experimental biologists, but it comes at the price of the integrity of systematics. It is not easy standing up to a mob, defending a discipline that dares to pursue excellence on its own terms, but we owe it to science, an ailing planet and future generations to do so. Humans have always been intrigued by morphological diversity and it remains among the most fascinating things about life. To downplay morphology is to lose the plot. Relegating it to footnotes and appendices in DNA-based studies eviscerates biodiversity exploration. Without detailed, accurate descriptions of morphology, DNA-based "trees" have very little of interest to explain.

Looking at a human and gorilla standing side by side, what we first notice are overall size, shape, posture, proportions, hairiness and so forth. In the moment, no one gazing at the pair sees percentages of difference in their DNA. Scientist or layperson, humans are visual creatures and what we immediately notice is morphology. It is morphology that yields a seemingly inexhaustible supply of improbable innovations and adaptations. How, we ask ourselves, is it possible to arrive at a bird of paradise, hummingbird and penguin beginning with a feathered, lizard-like creature resembling *Archaeopteryx*? Genetic differences are of interest, of course. But our problem-solving minds want to know the sequence of transformations responsible for such radically different looking species. Genes and developmental pathways complete the story, but morphology is critical to both the beginning and end of a journey to understand biodiversity.

Natural selection cares about morphology, too. How an organism is structured, how it functions in its environment, has all to do with whether it can survive, compete or attract a mate. Yes, genes store the information required for a developing embryo to end up looking like the species it is, but it is the physical body that interacts with its environment and that natural selection favors or culls.

Morphology is an example of the importance of emergent properties. Studying DNA, deconstructed into its constituent parts, becomes chemistry. At its most reductionist level it no longer contains information about biology, much less history. No study of the physical properties of individual nucleic acid bases can explain that which is most interesting about DNA. DNA gains emergent properties when bases are strung together. Likewise, emergent properties of morphology cannot be understood with knowledge limited to the molecular level. We already have the science of molecular genetics to tell us about that level of biological organization, and population genetics to inform us about phenomena within and between populations. We do not need systematics to mimic either of them. The aim of systematics is to explore and understand biodiversity *at and above the species level, at the granularity of individual characters*. It would serve science well to draw bright lines between these sciences to clarify responsibilities. Much of the 20th century was spent trying to redefine taxonomy in the image of other, more popular, sciences. Such efforts only made it less clear why we need systematics. It is time that systematics stands up for itself, clearly articulates its mission and ceases to be unduly influenced or overshadowed by priorities of other sciences.

As an aside, let's assume that we could understand the genome well enough to read what an organism looks like from its genes. This will never be completely possible because an organism's appearance may be influenced by injuries or infections, as well as epigenetics; that is, changes in gene function or expression due to environmental or behavioral factors. The spore-bearing surface of the crust fungus *Irpex lacteus*, depending on environmental conditions, may consist of many small holes, the pores that give polypore fungi their common name, or, alternatively, elongated tooth-like or spine-like projections. These forms are so dissimilar that they were assigned to different taxa in the past. Based on molecular data we could identify the fungus species and still not know whether it has pores or teeth. There will always be times when it is simpler, faster and more gratifying to simply look at the organism.

Laboratory mice, fruit flies, zebra fish, the mustard relative *Arabidopsis* and a few other model organisms have been relatively well described with respect to morphology. Many other species are, by comparison, only superficially described. And the morphology of at least 80% of species hasn't been described at all. In spite of tremendous progress since Linnaeus, the task of describing morphology remains in its infancy with major discoveries awaiting.

Some foolishly assume that taxonomists relied on morphology for the last 500 years only because they didn't yet have access to DNA. This is as preposterous as switching to a wholly DNA-based taxonomy. What we need is a taxonomic renaissance, a reawakening of curiosity about species and their attributes. Molecular data is a good partner, but a lousy substitute, for morphology.

Obvious in retrospect, the idea of homology, that structures can look very different yet be derived from the same ancestral body part, was a great conceptual advance for biology. Now, thanks to DNA and developmental studies, homology can be explored in greater detail enhancing the information content of morphology. How have developmental pathways been modified to result in diverse morphology? Restated by example, how have tissues of flowering plants been modified to produce fruits as diverse as tomatoes, kiwis and coconuts? Without molecular data we cannot fully explain the source of complex morphology, as evidenced by Minelli's study of body segmentation in myriapods. Without descriptions of morphology, there is no complexity to understand.

The *Biologia Centrali-Americana* is an impressive contribution to natural history. In it, more than 50,000 species from Mexico and Central America are recognized, in excess of 19,000 new to science. Published in London, between 1879 and 1915, the *Biologia* includes 63 over-sized volumes with 1,677 lithograph plates of which more than 900 are in color. It is a spectacular scientific, artistic and literary achievement.

One day, I walked into my laboratory and saw stacks of *Biologia* volumes. I asked my technician what he was doing with so many and was shocked when he told me that he had salvaged them from a dumpster behind the library. I was incensed that these icons of biodiversity would be trashed. After making some noise, I found myself standing beside the librarian in the dean's office. In her defense, these books are printed on acidic paper and slowly deteriorating, requiring sensitive handling by readers. She had received a "conservation grant" whose plan of work was to make photocopies on acid-free paper, then toss the originals. At least she had knifed the lithographs and filed them between sheets of tissue paper. All the same, it defied my understanding of the word conservation.

To my horror, I came to realize that our librarian saw herself as an information scientist. By all accounts, she had no particular affection for books beyond the words printed in them. On more than one occasion, I heard her paint a dreamy-eyed vision of a future time when libraries would contain no books, only computer-stored information. Some might say she was ahead of her time. I concluded that she was as mad as a box of frogs. Because *Biologia* volumes are such highly specialized books, they are rarely accessed by undergraduates and lovingly handled by faculty and graduate students who consult them. This means that manhandling was kept to a minimum and they were in no imminent danger of damage beyond repair. It was clear, even then, that we were on the cusp of major advances in digitization and that, within a few years, machine searchable text would become the norm.

Making photocopies was clearly an intermediate, and in this case unnecessary, step toward preserving knowledge printed on fragile paper.

To me, and millions of others, books are more than a primitive information storage system. They have historical, aesthetic and cultural significance beyond whatever facts and ideas they contain. There is something about the feel and smell of a book,[1] about holding someone's thoughts in your hands, and the power of ink on paper to form images in your mind and take you to unexpected places. I suppose it makes sense that if you are literally interested only in their content, you would see books as disposable.

This experience has given me insight into the thinking of those who would eagerly replace morphology with DNA. Their interest and curiosity about species stop at telling them apart, unlike the fascination with species and characters that has characterized taxonomy for centuries. With the limited goal of identifying species, of course they would see DNA as an acceptable shortcut to the information they seek and the careful description of morphology as an unnecessary waste of time. They would conclude that taxonomists had muddled along with morphology, as librarians had tolerated books, only because sequencers and electronic databases had not been invented. Had they been around in the 18th century, they think, we could have avoided all the hard work involved in comparing, describing and interpreting complex structures. It seems not to occur to them that morphology is interesting in its own right, that it uniquely tells us things worth knowing. Mark Twain quipped that there are things you can only learn by picking a cat up by its tail, and the same goes for morphology.

At first, it may appear contradictory that taxonomists describe what species look like and, at the same time, consider species and characters to be theoretical constructs. Although there are many species that are identifiable at a glance, species to a systematist are nonetheless hypotheses based on predictions about the distribution of characters. This is as it must be if species are to be testable scientific constructs. In many instances, it is necessary, too. It can be challenging to establish boundaries for widespread and variable species.

Beyond descriptions of what organisms look like, morphology is of great interest to systematists because of its historical information content. DNA is by comparison simple, consisting of various combinations of four bases—adenine (A), cytosine (C), guanine (G) and thymine (T). Properties of these bases are in the realm of chemistry, not biology. Any two guanine bases are indistinguishable anywhere in time or space. A single guanine base contains exactly zero historical information. This gets DNA's relevance to taxonomy off to a rough start. It is only by adding complexity that we begin to store information. Map the sequence of nucleobases in a strand of DNA and you add both complexity and the potential for retrievable information about history. Tease out sequences functioning as genes, you add information. Account for 3D folding of a protein, even more. Taking this idea of greater complexity being correlated with more information to its logical extreme, we can imagine an ideal data source that efficiently summarizes multiple genes, developmental pathways and thousands of nucleobases. There is a word for such an ideal data source: morphology.

While DNA can never replace morphology, the opposite is also true. Morphology is no match for DNA in certain respects. Species limits are not unambiguously established by degrees of DNA divergence, but such evidence can recognize variation that *might* indicate the presence of another species. Where dependable morphological characters are unavailable, DNA, judiciously used, can sort out likely species. DNA has revolutionized microbiology where many species cannot yet be cultured in the laboratory for close study.

It is strange to me that anyone could question the importance of describing morphology. Who is not intrigued by the bioluminescent esca of a deep-sea anglerfish or the buttresses of a strangler fig? The entire point of taxonomy, done for its own sake, is to get to know species and their characters and understand them in an evolutionary-historical context. The first thing we typically notice about a species is its form: centipedes, cycads and canaries don't look much alike. Anyone who doesn't see the morphological difference between a sardine and shark had best stay out of the water. If all species looked identical, there would have been little impetus to search for a theory of evolution, or cause for celebration when one was found.

It does not take an especially long sequence of letters to uniquely identify a book. For example, these are the first 50 letters, in order, on page one of two books:

proceedsolinustoprocuremyfallandbythedoomofdeathen
byanyobjectivemeasuretheamountofsignificantoftentrau

Given stacks of copies of the two books, we could efficiently sort them into two piles of the same title by these 50 letters alone. But this rather misses the point of books. It robs us of their information content and mutes the authors' voices. Separating these letters into words with spaces and punctuation, and reading them in the order intended, we come away with a more rewarding experience. A molecular-based taxonomy offers similar sorting as a replacement for information-rich morphology.

The two books, by the way, are Shakespeare's *The Comedy of Errors* and John Kotter's *Leading Change*. While I found the latter useful in my roles as an institutional leader, and with apologies to Mr. Kotter, I would not rank it alongside the Bard as a great work of Western literature. Identifying books and telling them apart are of little use or interest unless we are also able to engage with the ideas they contain and appreciate their contents as individual creations. The same is true of species. Identifying them, simply telling them apart, is necessary for biological fieldwork, intercepting pests at ports-of-entry and so forth, but insufficient if we wish to understand species or their evolutionary story. Much intriguing information is found in complex characters, especially those of morphology. A DNA-based taxonomy is purely utilitarian, aimed at meeting bare minimum needs of other biologists to identify species while ignoring the mission of systematics as a basic science. DNA data excels at quite a few things, but exploring, documenting and understanding species are not among them, at least not when DNA is used alone. DNA is also useful, but not essential, for analyzing phylogenetic relationships. But analyzing phylogenetic relationships without knowledge of morphology and natural history is like reading a novel without a plot.

DNA barcodes, short snippets of DNA used to tell species apart, are analogous to UPC codes on grocery items and about as relevant to taxonomy's mission. They only identify already-known species on a planet where most species are unknown. Unless authenticated sequences exist for species that are known and corroborated, there is no credible benchmark to compare them to. Compared among themselves, sequences of any length are merely descriptions of genetic diversity of unknown significance. It is possible to assume that any population separated from others by the average distance between known species, or more, signals a species discovery, but this is a gamble. There are no inviolate rules regarding degree of difference and species status. So, when it comes to using metagenomics to assess the diversity of DNA in an environment, I'm not sure what it is, but it isn't taxonomy.

Many molecular systematists have focused on reanalyzing relationships among already-known species whose morphology was described in the past. Howls of triumph erupt and bony fingers point accusingly when DNA evidence contradicts traditional ideas of relationships based on morphology. It is hailed as a vindication of molecular data and indictment of morphology. We are told that when morphology and DNA tell conflicting stories, DNA is to be believed. But, why? Because it's new? Because it uses expensive equipment rather than careful observations and disciplined reasoning?

In reality, all that we can say is that one or the other, or both, have been misinterpreted. Because there is only one evolutionary history, all relevant data properly analyzed must conform to one and the same pattern. In the 19th century, Louis Agassiz spoke of the three-part parallelism among morphology, fossils and developing embryos, all pointing to the same pattern that we now attribute to common descent. To this we add molecular data but should remember that it is just that: one source of data, no better or more reliable than any other.

If the hubris of molecular systematists looks familiar, it is. Gareth Nelson reminded us of paleontologists in the early 20th century who believed the fossil record *revealed* the true evolutionary history. In reality, the stratigraphic record must be carefully interpreted. The same is true of DNA. Sequence data is relevant to phylogeny hypothesized, not phylogeny revealed. Conflict with morphological data is a warning that *both* data sets are suspect until we determine what went wrong in the analysis of at least one of them. There are interesting things to be learned when such conflict arises—perhaps an insertion in a sequence, or convergent evolution of similar structures—that explain why the initial interpretation was mistaken. While conflict indicates a problem, it's important to recognize that concordance is expected because genes, developmental sequences (i.e., ontogeny) and morphology are not biologically independent of one another.

Do you remember the last time you visited a zoo? Imagine that every animal on display looked absolutely identical with the exception of DNA profiles posted on placards. There would be few photos taken or repeat visits to such monotonous menageries. It is the diversity of morphology that is on display, and that draws us back. *Vive la difference!* The cave paintings at Lascaux are evidence that our ancestors paid attention to morphology, too. Pondering the meaning of the pattern of similarities and differences among species is as old as humankind. Among the first spoken words were almost certainly names for commonly encountered species. Phrases like, "Watch out for that mastodon!" and "Don't eat that red berry!" were a matter of life and death. Without DNA sequencers, our forebears found it useful to know enough morphology to tell species apart on sight. Even with them, it still is.

Without morphological diversity, would we bother to study phylogeny? What, exactly, would we be trying to explain? Systematics came about as a scientific way to explore and understand the diverse attributes of species. As Gareth Nelson and Norman Platnick pointed out, it was the accumulation of taxonomic descriptions, and an obvious pattern of similarities and differences among species, that begged to be explained. Without careful comparative and descriptive morphology carried out by taxonomists, there would have been no predicate for searching for causal processes. As they succinctly put it: no pattern, nothing for Darwin to explain.

It is intellectually satisfying to survey morphological diversity among species and, following careful analysis, understand how the parts relate to the whole story. What had seemed utter chaos suddenly falls into place. Diagnosing species and analyzing their relationships with DNA may meet the narrow need of biologists to identify species and know something of relationships, but it is bare minimum knowledge and we deserve more.

In the early years of DNA sequencing, I recall a bizarre conversation with a graduate student. Animated by his excitement, he proclaimed that DNA was "new" data and, to him at least, therefore superior to "old" data like morphology. Aside from the inconvenient fact that morphological structures are exactly the same age as genes coding for them, why should we assume that new is better than old? Why engage in scientific research at all unless we occasionally get things right? That Galileo's observation of moons around Jupiter was made in the early 17th century in no way detracts from its significance or makes it any less true. There is a reason that new, better and correct are not synonyms.

We perceive an elephant's trunk as unique only because deer flies, dogs and daffodils lack one. Whether an attribute is considered unique depends upon knowledge of other species. Such knowledge is recorded in taxonomic descriptions and it is the comparative method of systematics that uncovers patterns among species, the same pattern that leads us on a journey of discovery about relationships and the chronological order of morphological transformations.

There are so many insects that entomologists have always been called upon to make identifications, from agricultural pests to disease vectors and unwelcome household guests. As a student, I was offended by colleagues who, for that reason, mistakenly assumed that this is why taxonomy exists. It is ironic that now, with the advent of DNA data, taxonomy really is in danger of becoming the mere service it was long and falsely held to be. I do not mean to detract from the noble practice of applied taxonomy, but to argue for the fundamental research that makes it possible.

At present DNA has many champions and the lion's share of positions and funding, so I feel no need to apologize for defending descriptive taxonomy. The reason I became a systematist was my fascination with morphological diversity. To minimize that which is most interesting for purposes of identification expediency is a huge mistake. Taxonomists have always done their duty by making species known, named and identifiable, and they will continue to do so using the full range of data and technologies available to them. But this in no way diminishes the importance of their fundamental research.

Comparative morphology, like Linnaeus' dream of a worldwide species inventory, was ahead of its time. In the early days of taxonomy, it was difficult to reach museums, remote collecting sites and colleagues to study, collect or discuss specimens. And publishing illustrations of morphology was prohibitively expensive. Advances in communication, travel, digital imaging and publishing have changed all of that. Specimens scattered among museums around the world have never been more accessible. Web-based publications offer all the space needed for images of morphology at virtually no cost. And digital imaging has come into its own, from high-resolution light microscopy to phase and differential interference contrast, confocal, composite 2D and 3D images, computer-assisted tomography and scanning electron microscopy. With millions of unknown species, each with a bounty of morphological characters, the time for a modernized, reinvigorated morphology has arrived. And given the extinction rate, just in the nick of time.

Morphology as done by taxonomists is at a scale that falls short of the expectations of a pure morphologist. Robert Snodgrass, who wrote the book on insect morphology, explained that the pure morphologist wants to understand a structure in full detail. This may involve dissections, observations of developing embryos, sectioning and fine structure studies and experiments to determine function among other things. This is to say that typical descriptions in botany and zoology are sufficient for the purposes of taxonomy but unsuited to certain others. There may be times when a deeper dive into morphology is necessary to interpret a character, but to make such in-depth morphology a requirement of taxonomy would in most cases be uncalled for and would slow the description of species. There is a commonsense middle ground for documenting species. Molecules and morphology each occupy a particular place among a panoply of data and neither should claim to be a suitable substitute for all others.

Mastering morphology, even at the scale of descriptive taxonomy, is no simple thing. Knowledge of scores to hundreds of structures and the precise terms for them is just the beginning. Interpreting complex morphology demands deep thought about details of structure and hypotheses of homology. This accounts for the surprising amount of knowledge, thought and theory behind a competent description. Just describing what an organism looks like, or sequencing a strand of DNA, reminds me of something Charles Darwin said about geology practiced without deep thought, quoted by Michael Ruse: "a man might as well go into a gravel-pit and count the pebbles and describe the colours."

There have been embarrassing examples of molecular genetics practiced with limited knowledge of the species involved. There was a graduate student who spent several years sequencing what turned out to be the wrong species for his thesis. The aim was to sequence a particular fungus species, but it turned out that he had been sequencing a mold contaminant in his cultures. Because he had never bothered to learn to recognize the target fungus on sight, he had no way of knowing his project had gone off the rails.

Another way of looking at morphology is quality versus quantity. We should aim for more knowledge, not more data. The genomes of humans and gorillas differ by less than 2%, so sequencing every one of 3.2 billion base pairs is an overkill for distinguishing the two. Most of this data has nothing to say about how humans and gorillas differ, nor much about their phylogenetic closeness. Humans, chimps and gorillas all share a common ancestor and have only fractionally diverged in genome and morphology. More evidence of shared, primitive DNA similarity tells us nothing about sister species, merely the lack of divergence from the ancestral genome. What matters is the opposite: shared-derived similarities, or synapomorphies.

Evolutionists have long dreamed of identifying biological laws as reliable as those in physics, but this search is futile. Evolution is too complex, too contingent. The Belgian paleontologist Louis Dollo proposed one such law in 1893, often stated as "a complex structure lost in evolution cannot be regained in its same form." This holds, Dollo opined, because a structure "always keeps some trace of the intermediate stages through which it has passed." Such a trail of bread crumbs is a consequence of complexity. A single complex character can provide extremely compelling evidence of relationship. Presented with an African elephant, Asian elephant and white rhinoceros, genome sequencing is hardly necessary to surmise relationships.

We have mentioned that analyzing multiple sources of data is useful because incongruent patterns alert us to mistakes in interpretations. Complex characters, like those of morphology, have the added benefit of being internally testable, too. It is possible to reexamine them in greater detail: form, position, constituent parts, microstructure, embryonic development and DNA sequences. Morphology-based taxonomy has a tradition of carefully assessing individual characters. Molecular data and computer programs have fostered a different approach, analyzing an entire data matrix at once with little attention to individual characters. This gets to the shortest branching path for the tree that then only needs to be rooted to be seen as a phylogeny. When the goal is to understand characters, however, there is enormous advantage in analyzing them individually.

It is worth noting that accurate descriptions of morphology have an inordinately long shelf life. Excellent morphology-based descriptions have no expiration date. A taxonomic monograph, done to high standards, remains useful for centuries. It will be deficient in respect to recently discovered species or characters, of course, yet may remain accurate in its characterizations. Given the number and magnitude of changes in the theoretical *milieu* of biology over the period of modern taxonomy, this timeless quality of careful morphology is truly impressive. Few areas of research yield such long-enduring results.

Museum specimens are morphology's analog to repeated experiments. Questioned characters can be revisited by studying the actual specimens previously described. And if questions remain unresolved, there are recourses to better understand morphological structures. David Sharp and Frederick Muir, in their classic study of the male genitalia of beetles, cited a quotation by Edward Gibbon who demonstrated his awe of morphology when he compared a bug with the cathedral of Saint Sophia at Constantinople:

> the enthusiast who entered the dome of St. Sophia might be tempted to suppose that it was the residence, or even the workmanship, of the Deity. Yet how dull is the artifice, how insignificant is the labour, if it be compared with the formation of the vilest insect that crawls upon the surface of the temple!

We can only gain full appreciation for morphological diversity and evolutionary history if we collect species before they are gone.

Note

1 With a particular weakness for vanilla cakes and ice cream, I was delighted to learn that there is science behind my love of old books. The International League for Antiquarian Booksellers report that a common odor of old books is a hint of vanilla: "Lignin, which is present in all wood-based paper, is closely related to vanillin. As it breaks down, the lignin grants old books that faint vanilla scent." Colin Schultz discusses this in his article in *Smithsonian Magazine*: smithsonianmag.com, 18 June 2013.

Further Reading

Callaway, E. (2021) Million-year-old mammoth genes shatter record for oldest ancient DNA. *Nature*, 590: 537–538.

Crowson, R. (1970) *Classification and Biology*, Heinemann, London, 350 pp.

Dollo, L. (1893) The laws of evolution. *Bulletin de la Société Belge de Géologie*, 7: 164–166.

Gibbon, E. (1788) *The History of the Decline and Fall of the Roman Empire*, Strahan and Cadell, London, Vol. IV, Ch. XL, p. 248.

Hennig, W. (1966) *Phylogenetic Systematics*, University of Illinois Press, Urbana, 263 pp.

Hennig, W. (1981) *Insect Phylogeny*, John Wiley and Sons, New York, 514 pp.

Luc, M., Doucet, M. E., Fortuner, R., et al. (2010) Usefulness of morphological data for the study of nematode biodiversity. *Nematology*, 12: 495–504.

Mathew, W. D. (1925) Recent progress and trends in vertebrate paleontology. In *Annual Report of the Board of Regents of the Smithsonian Institution for the Year Ending June 30, 1923*, Smithsonian Institution, Washington, 273–289.

Minelli, A. (2020) Arthropod segments and segmentation—Lessons from myriapods, and open questions. *Opuscula Zoologica (Budapest)*, 51 (supplement 2): 7–21.

Nelson, G. (2004) Cladistics: Its arrested development. In *Milestones in Systematics* (eds. D. L. Williams and P. L. Forey), CRC Press, Boca Raton, pp. 127–147.

Nelson, G. and Platnick, N. (1981) *Systematics and Biogeography: Cladistics and Vicariance*, Columbia University Press, New York, 567 pp.

Owen, R. (1894) *The Life of Richard Owen*, Vol. I, John Murray, London, 409 pp.

Platnick, N. (1978) The transformation of cladism. *Systematic Zoology*, 28: 537–546.

Rupke, N. A. (1994) *Richard Owen: Victorian Naturalist*. Yale University Press, New Haven, 480 pp.

Ruse, M. (1996) *Monad to Man: The Concept of Progress in Evolutionary Biology*, Harvard University Press, Cambridge, 628 pp.
Santos, L. M. and Faria, L. R. R. (2011) The taxonomy's new clothes: A little more about the DNA-based taxonomy. *Zootaxa*, 3025: 66–68.
Sharp, D. and Muir, F. A. G. (1912) The comparative anatomy of the male genital tube in Coleoptera. *Transactions of the Entomological Society of London*, 60: 477–642.
Snodgrass, R. E. (1935) *Principles of Insect Morphology*, McGraw-Hill, New York, 667 pp.
Snodgrass, R. E. (1951) Anatomy and morphology. *Journal of the New York Entomological Society*, 59: 71–73.
van der Valk, T., Pecnerová, P., Díez-del-Molino, D., et al. (2021) Million-year-old DNA sheds light on the genomic history of mammoths. *Nature*, 591: 265–269.
Wheeler, Q. (2008) Undisciplined thinking: Morphology and Hennig's unfinished revolution. *Systematic Entomology*, 33: 2–7.

12 The Inventory Imperative

As clouds of extinction gather, we ignore perilous deficiencies in what we know. Unprepared for the environmental storm that lies ahead we fly blindly, without navigation charts or a clear destination. Unless we learn the species that exist, we cannot monitor extinctions, assess the integrity of ecosystems, set measurable conservation goals or explore their origins. As millions of species race toward extinction, we are neither organizing an inventory nor repairing damages done to taxonomy by years of neglect. Distracted by scientific fads and fashions, we live with a rate of species discovery essentially unchanged in nearly a century. The only obstacle to making an account of earth species is our lack of resolve.

It is conceivable that you will hear more about Chupacabras, Big Foots—or is that Big Feet?—and Roswell Greys than any of thousands of actual new species discovered this year. Quietly, out of sight, taxonomists continue their centuries-old hunt for species. The pace of naming new species was on the rise in the early 20th century, reaching an annual high-water mark of about 20,000 just before World War II. Following a wartime hiatus, this rate returned—but not the upward trajectory. Annual species discoveries have remained more or less constant since. Considering the expansion of academia, airline routes, science funding and research tools since 1945, and the low-hanging fruit of millions of undescribed species, this lack of increase in the rate of discovery is due to misplaced priorities rather than any external limitation.

We are engaged in a mission to find evidence that we are not alone in the Universe. From the Search for Extraterrestrial Intelligence (SETI) eavesdropping for messages from outer space to soil-sampling Martian rovers and sky-scanning telescopes, the search for extraterrestrials (ETs) is on. One result is a growing list of earthlike planets, worlds in so-called Goldilocks zones around stars—not too close and hot, not too distant and cold, but just right for life. Human nature being what it is, the grass looks greener on the other side of the galaxy. Drawn by the allure of extraterrestrial life, we shamefully fail to appreciate the astonishing life forms around us. Discovering living things in the Milky Way should be a big deal, whether found on another world or this one. But knowing enough about species to maintain earth as a hospitable planet ought to be a top priority. Surrounded by a profusion of species, we thoughtlessly take them for granted. Regardless of whether extraterrestrial life

DOI: 10.4324/9781003389071-13

forms are eventually discovered, earth species will remain special by virtue of their contributions to the biosphere and our shared evolutionary history.

Are we alone in the Universe? Absolutely not. It is true that there are no verified visitations by flying saucers or hard evidence of extraterrestrial life. But we know with absolute certainty that millions of life forms exist in one corner of the Milky Way because we share their world. Unlike the mere possibility of locating life elsewhere, exploring biodiversity on earth is a sure thing. It will pay rich dividends in knowledge, improving and enriching our lives. And as a bonus, travel to explore species here has regularly scheduled flights and is considerably more comfortable than the cramped quarters of a space ship.

I personally believe that extraterrestrial life must exist and I would love to be around when it is discovered. It will be a milestone in human history but should not come as a surprise. Its greatest impact may be on religion rather than science, although both will no doubt have revisions to make in their thinking. As we evaluate priorities, we must not lose sight of the fact that we have an eternity in which to seek out life on other worlds, but just one chance to explore the diversity and history of life on earth. Let's not miss it.

If and when we do come face to face with ETs, it may be on a planet much like the earth was billions of years ago, with warm shallow seas and no one home except single-celled microbes. What makes earth life most worth exploring is its incredible diversity, unlikely evolutionary story, and what it can teach us about ourselves. To pass up the last opportunity to explore planet earth's species in order to search for life elsewhere, life that may or may not exist, is madness.

Assuming that we one day find another biologically diverse planet, engineer space craft capable of taking us there and complete an inventory of its species, we will next want to know how its species diversity, evolutionary processes and history compare to those of earth. If your primary interest is exobiology, then you should demand support for an inventory of earthlings so that we are ready to compare and contrast whatever we discover out there. It will be a scientific disgrace if we ultimately know more about the diversity and history of life on some world other than our own.

A *Taxonomic* Inventory

Metagenomics, isolating and sequencing DNA from environmental samples, is being hailed as a rapid and efficient way to discover species without the need to educate or support taxon experts, or build and maintain collections. But if remote glimpses of what may or may not be actual species is an acceptable substitute for descriptions and scientific hypotheses, why not credit the first photograph of earth from space, made aboard a German V-2 rocket in October, 1946, with the discovery of millions of species? They are, after all, in the picture. Setting aside the fact that exactly what a microbial species is remains unsettled, metagenomic surveys measure genetic diversity, not species diversity. Until we know more about both, this will remain so. Thus, even if we only want to know how many species exist, rather than *what* species exist, metagenomics is not up to the task. Because we have only one chance to explore biodiversity, we need to learn as much as we can about species, not DNA surrogates.

A simple inventory could consist of a list of species names or DNA sequences, but that is not my idea of an inventory. I refer to a *taxonomic* inventory with the goal of documenting and learning all that we can. The spirit of such an inventory was expressed in the report from a workshop that I organized in New York:

> Our goal is no less than a full knowledge-base of the biological diversity on our planet, by which we mean: knowledge of all Earth's species, and how they resemble and differ from each other (i.e. all their characters from detailed morphology to as much genomic information as is feasible to collect); a predictive classification of all these species, based on their interrelationships as inferred from all these characters; knowledge of all the places at which each of these species has been found with as much ecological data as are available from specimens in the world's collections (e.g. host data, microhabitat data, phenology, etc.); and cyberinfrastructure to enable the identification of newly found specimens (including automated identification systems based on images and genomic information), the efficient description of species, and open access to data, information and knowledge of all species by anyone, amateur or professional, anywhere, any time.
>
> (Wheeler et al., 2012)

While such an inventory is structured around the aims of systematics, it is not incompatible with the interests of ecology, biogeography and conservation. Geographic, temporal and ecological data associated with specimens are all recoverable from a database. As extinction progresses, such data will become increasingly valuable. While classifications and the organization of specimens in natural history museums should reflect phylogenetic relationships, species may be virtually reassembled into communities based on environmental requirements, locality data, dates and co-occurrences to provide insights into the combinations of species found in ecosystems prior to the great extinction.

Such an ambitious, information-rich inventory cannot succeed unless systematics succeeds, too. Investments in systematics must address the workforce, including professional and amateur taxonomists, modernization of research infrastructure and an unprecedented expansion of natural history museums. Unless museums accept the challenge of becoming our planet's memory of biodiversity and biosphere, science and society will suffer.

Return on Investment

Benefits derived from a complete species inventory are too numerous to list, no doubt including many not yet thought of. Any one of the following rewards would, by itself, easily justify the cost of a global inventory.

Biodiversity Baseline. Habitats are changing at a pace unseen in human history. As a consequence, we confront environmental challenges unprecedented in kind, scale and complexity. We know species are going extinct but have no way to detect which, how many, where or when. As ecosystems are degraded and fragmented, we can't be sure which or how many species are affected. Similarly, as climate change shifts geographic and seasonal distributions of species, we are uncertain of the impacts. And, without a catalogue of native species, it is difficult to recognize invasive species that threaten crops, livestock and human health.

Whether biodiversity is stable, decreasing or increasing for a particular location can only be determined by benchmarking it against which and how many species existed there to begin with. Without baseline information about the species present in an ecosystem, how are changes in the web of life to be recognized? It is irresponsible to make decisions about conservation and land management without baseline knowledge of species.

Deep Ecology. Human activities are driving species extinction, accelerating the disruption of our planet's ecosystems. These systems are the source of essential ecological services and the engines that perpetuate life. Each decision we make about land development, resource exploitation or conservation influences the outcome of environmental changes already underway. Unaware of the species present in an ecosystem, we measure its functions, keep track of a few key species and simply hope for a good outcome for the rest of its inhabitants. We can do much better.

This puts a fine point on an ill-advised trend in ecology. In recent decades, ecology has increasingly focused on ecosystem functions with little regard for the combinations of species that differentiate such systems from one another around the world. Taking this functional bias to its logical extreme, we can imagine bioengineering super trees that do the work of scores of naturally occurring plant species capturing solar energy, sequestering carbon, releasing oxygen, cycling nutrients and purifying water. So long as there are the same levels of in-puts and out-puts in a system, who cares which or how many species are involved? People curious about the composition and origins of the biosphere should care. So should those wishing to see ecosystems remain sufficiently diverse to be resilient in the face of changing circumstances.

When I was director of environmental programs at the National Science Foundation I would, from time to time, have a pint with one of my ecosystem program officers and discuss our views on science. He was amused by my intense interest in species. He took a 30,000-foot view of the planet and was concerned with carbon budgets. What mattered to his world view was how much carbon was tied up in forests and frozen tundra, and how much was being released into the atmosphere. Which or how many species lived in the places where these planetary functions were carried out was an irrelevant detail to him. I was equally astonished that anyone could fail to appreciate the majesty of biodiversity and evolutionary history, only visible at the granularity of species and characters.

The old science fiction dream of a permanent human colony on Mars is now being pursued seriously. In fact, preparations for a manned flight to Mars are underway. If we can build a sustainable village under a dome on Mars, through some combination of engineering and plant cultivation, then clearly it could be done on earth more easily and cheaply. It is not evident that a high level of biodiversity is necessary to provide the minimum ecosystem services required to support human life. Seriously degraded or artificial ecosystems might be good enough. Ecology's single-minded obsession with functions could eventually backfire, providing unintended rationale for turning our backs on the preservation of biodiversity. Before we permit millions of species to disappear, we must ask whether there are reasons to value and protect them beyond the services they provide.

I don't question that humans could survive under a colossal cloche on Mars but would that really be living? Speaking for myself, I think not. Peering out across a red, barren solar-wind swept landscape, I would feel a smothering sense of isolation and loneliness—no matter how many people or potted plants were around me. We are, by virtue of our evolutionary origins, a part of, not apart from, biodiversity. While enjoying family, friends and colleagues, I take a special comfort in the company of other species.

The first time I set foot in a lowland tropical rainforest was as a graduate student visiting the Smithsonian research station on Barro Colorado Island in Panama. From my very first steps down a muddy trail into the forest, I experienced an overpowering sensation of coming home. Although surrounded by plants and animals I had never seen before, I felt a deep sense of kinship, of belonging. This is the hardwired part of our psychological makeup that E. O. Wilson called biophilia, a product of the evolution of our brain discussed by Carl Sagan in *The Dragons of Eden*. Just as a child feels secure in the arms of parents who have been ever present in her life, humans are subconsciously comforted by knowledge that nature can be relied upon to be there, whatever else may come. To cavalierly destroy ecosystems and permit species to go extinct is to risk being set adrift alone in a very hostile Universe, with no safe harbor to which we can retreat.

Although rarely considered in process-obsessed ecology research, ecosystems have a fascinating history, too. The systems we see today are the result of a complex series of geologic, climatic, dispersal, extinction and speciation events. Each species added to or subtracted from a system has an impact, large or small. Even the least impacts, aggregated in numbers, can become existential threats to the integrity of the system as a whole and its capacity to deliver services. In short, it matters which and how many species coinhabit an ecosystem and no combination of assumptions, guesses or computer models can substitute for actual knowledge of species and their interactions.

Because taxonomy has been ignored, ecologists have had to make do with scant information about species, particularly in the most species-rich regions of the globe. Rather than rallying to taxonomy's cause and advocating funding for species exploration, ecologists have pursued workarounds from rough-sorted "morphospecies" (grouping similar specimens into provisional "species," labeled A, B, C, ...) to DNA barcodes. They regard taxonomy as foreign because it is nonexperimental, historical and focused on patterns rather than processes. This intellectual xenophobia is ultimately injurious to ecology itself. Deep understanding of ecosystems depends on the ability to drill down, when appropriate, to the level of species-to-species interactions. That, of course, is possible only when taxonomists are supported to do their job, determining what species exist and making them recognizable. Embracing quick-and-dirty species identification schemes in place of rigorous species hypotheses treads on thin ice.

It is easy to understand why an ecologist would simply want to identify species as easily as possible. Combine this with the misconception that species are arbitrary, and shortcuts begin to make sense. But with ecological research and conservation decisions carrying irreversible impacts on biodiversity, we are better served when systematics is carried out to high standards of excellence. Happily, supporting systematists to reach their goals also results in the most, and most reliable, information.

The ability to identify species opens doors for ecologists. In return, knowledge of ecology helps systematists better understand species. But the sciences differ fundamentally in goals and methods. Taxonomists recognize the importance of supporting ecology done on its own terms. But, for whatever reasons, there are continuing efforts to force systematics into the mold of experimental biology. Essential to the environmental sciences, systematics is nonetheless an evolutionary science focused on historical patterns. In contrast, ecology is experimental, looking to explain how organisms interact and ecosystems work. There are important exchanges of knowledge between the two, but the quality of knowledge depends on each science's autonomy. Ecology does not exist to explain to taxonomists what species do any more than systematics exists to identify species for ecologists. Yet both fields are necessary if we are to understand biodiversity and respond effectively to the current crisis.

Now, more than ever, we need to cultivate what should be a natural alliance between ecology and systematics. This requires each to acknowledge their differences and have mutual respect for each other's contributions. We need deeper understanding of the biosphere, including its evolutionary origins and present-day functions. It is as tragically misguided to shoehorn taxonomy into the framework of ecology or genetics, as it would be to neglect them in a rush to complete a taxonomic inventory. It is for ecologists to determine how deep they must go to fully understand systems, but only complete and reliable knowledge of species gives them that option.

Evidence of Evolutionary History. The fossil record provides unique insights into evolutionary history but, counterintuitively, is less informative than comparative studies of extant species. There are, of course, species, characters and character combinations observable in the fossil record that no longer exist among the living. And fossils can inform us of the minimum age of lineages, based on their earliest occurrence in the geologic record. But fossils preserve only a small fraction of informative characters, most often hard parts like bones and shells. For this reason, specimens of soon-to-be-extinct species that we preserve in natural history museums will be, for practical purposes, perfectly preserved "fossils" from which we can learn an extraordinary amount.

We cannot predict which or how many species will ultimately go extinct. Therefore, science is served best by making natural history museum collections representative of all species, a comprehensive mirror on biodiversity as it exists in the 21st century. From such museum specimens, systematists can continue to refine our understanding of phylogeny, verify places where extinct species once occurred, explore morphological evolution and illuminate clues for biomimetic designers and engineers.

Measurable Conservation Goals. In cynical moments, I see many conservation projects as insincere window dressing. The focus is often on saving a few showy plants and charismatic mammals and birds, rather than conserving the diversity of life. As the argument goes, if we set aside places where such "sexy" species live, others will come along for the ride. This may or may not be true, and I hope that it is. But either way, we can only be certain that the masses of species are being saved if we are able to recognize and monitor them.

About the same time that there was handwringing over the status of the spotted owl, *Strix occidentalis*, on the Pacific coast, Pedro Wygodzinsky of the American Museum of Natural History discovered a small, inconspicuous primitively wingless insect in the California coastal forest. While the population of the northern spotted owl was indeed threatened, the species as a whole was not. Other populations existed in spots (no pun intended) as far away as Mexico. In contrast, Wygodzinsky's discovery, *Tricholepidion gertschi*, is a remarkable "living fossil." It belongs to a family, Lepidotrichidae, previously known only from fossils and thought to have been extinct for tens of millions of years. What makes it even more interesting is its morphology that appears little changed from that of its ancestors, and its position in the tree of life. It belongs to a lineage closely related to Pterygota, the major branch that includes hundreds of thousands of species of winged insects. To say it crassly, conserving this little silverfish, the last living species of an entire ancient family, offering insights about a species very similar to the ancestor of all winged insects, is of greater scientific importance than one population of spotted owl. I, of course, prefer that both be conserved. And I'm not naïve enough to think that a silverfish could compete with a cute, big-eyed, fluffy owl as a poster child for conserving California's coastal forest.

Stating the obvious, only species surviving the current extinction event will be available for future evolution. Being a selfish species, humans have so far given this fact little weight. Distant evolution may happen after humans have killed themselves off or emigrated to other worlds. And, in any event, no one living today will be around to benefit from future evolution. We may not witness it ourselves, but we have an ethical obligation to give it serious consideration in our planning.

Some scientists, including Terry Erwin, have argued that we should favor the survival of recently evolved species. They reason that because such species have recently evolved themselves that they are likely to continue to mutate and spin off new species. I take a different view. Many recently evolved species are highly specialized, finely tuned for life in a narrow niche. In contrast, species like Wygodzinsky's silverfish are generalists. In this case, a morphologically conservative detritivore that may have a similarly conserved genome. If so, then its genetic makeup may resemble that of an ancestor that was omnipotent enough to give rise to hundreds of thousands of diverse descendant species. Just the thing that will be needed to fill niches vacated by millions of extinctions.

Such considerations cast a different light on our response to the biodiversity crisis. It is not enough to worry about ecological services and our short-term needs. Surely there is room on the ark for species that contribute to morphological, ecological and genetic diversity, too. If we wish to give evolution and civilization as many options as possible in the future, then we must learn about species diversity now.

Biomimicry Idea Bank. Biomimicry is not new. Humans have always taken notice of other species and mimicked them in various ways. Organized biomimicry, however, is a recent phenomenon. Janine Benyus has brought attention to the possibilities, elevating biomimicry as a way of thinking and rich source of ideas for problem-solving. Following her lead, it is time for systematists to do their part on the supply side of a biomimicry revolution. Knowledge created by systematists can transform biomimicry by opening a superabundance of models. Approached in the right way, a species inventory can unveil attributes of millions of species with the potential to inspire biomimetic innovation. Next generation biomimicry will come from a strategic partnership among taxonomists, information scientists, engineers, designers, inventors, economists and entrepreneurs. Descriptive taxonomy is key to a nearly inexhaustible supply of inspiring ideas.

Natural Resources. Ethnobiologists tell us that human civilizations use, or have used, tens of thousands of species for food, fiber, medicines and other purposes. If the ratio of "useful" to total species holds—to date, about 50,000 out of 2 million species—then we can reasonably expect to discover another 200,000 species of utility to us. This, however, is a lower estimate. First, because so few of the 2 million named species have been seriously examined for what they might be used for. And second, because a biomimetic awakening promises to broaden our perception of useful species and natural resources.

Language of Biodiversity

When taxonomists name species and higher taxa, they create a language of biodiversity. With millions of species, it is essential to have a system of informative, stable names with which to store and communicate information. In this language of biodiversity, scientific names constitute the vocabulary and theories of phylogenetic relationship the syntax.

I'm not sure about you, but I have no idea how to remember, much less pronounce, a DNA sequence like ATGCTTCAGGCTA. I have enough trouble keeping track of PIN numbers and passwords. Short DNA sequences (i.e., "barcodes" a few hundred base pairs in length) can be used to uniquely identify species, leading some zealots to suggest doing away with Linnaean names altogether. DNA barcodes or arbitrary numbers may be sufficient for computer-based information storage and retrieval, but we have come to expect more from scientific names. They allow us to speak to one another with precision. And when names are descriptive, they are easier to remember.

Many biologists spend their entire career studying one or a few species in the laboratory or field, so limitations of barcodes or numeric identifiers are less apparent to them than to the taxonomist who routinely deals with hundreds of species. The importance of names, however, is deeper. It is innate to our psyche, part of being human.

Linnaean names are as stable as we need them to be. Biologists complain bitterly when taxonomists change the names of species, but this is the price for names that are accurate, informative and up to date. Scientific names at all levels are associated with hypotheses that are subject to change in response to new evidence. When species hypotheses are tested and rejected, names associated with them must change to keep pace. When what was believed to be one species is later found to be a complex of species, its name must be narrowed to refer to just one among several. Or, if what was thought to be a species turns out to be no more than a variant within another species, its name must be subsumed in synonymy. Eugene Gaffney put it best when he said that taxonomic stability is ignorance. As knowledge grows and improves, we want a system of names capable of changing to reflect it.

It is impossible to overstate the importance of language for humans. As Steven Pinker put it, "Humans are so innately hardwired for language that they can no more suppress their ability to learn and use language than they can suppress the instinct to pull a hand back from a hot surface." Given this irrepressible instinct and capacity to communicate with language, why fight it? It makes more sense to embrace this remarkable human capacity to use words as we explore and document millions of species instead of turning to less intuitive, arbitrary species designators.

Existing names for species and higher taxa already constitute a "dictionary" with millions of entries. Arranging species into higher taxa, groups are given additional names with their own meanings. Without this language, it would be difficult for scientists to communicate. Advocates of the "phylocode," a poorly conceived plan to replace the Linnaean system, undervalue the role of language in systematics. They ignore the fact that the Linnaean system has proven itself by giving voice to a series of paradigms in biology—from Creationism to Darwinism and phylogenetic systematics—all while remaining equally informative and useful. They propose that we use mononomials, that is, single words, as species names. For existing species, they would simply collapse genus name and specific epithet into a one-word name. Thus, the human species *Homo sapiens* becomes *homosapiens*. Meanings of names would be more difficult to decipher and pronunciations more challenging. And, if you think the binomial for the Southeast Asian soldier fly, *Parastratiosphecomyia stratiosphecomyioides*, is a mouthful, try *parastratiosphecomyiastratiosphecomyioides*. At least the Linnaean version lets you come up for air.

Linnaeus' convention of using two words for a species name is pure genius and should be retained. His binomials, by incorporating names of genera, tell us something of relationships and permit us to use the same descriptive words for species over and over again. Thus, the European silver fir, *Abies alba*, and the white oak, *Quercus alba*, both use the epithet *alba* that means white. Such epithets are not always descriptive, but when they are, it is an aid in both recognizing a species and remembering its name.

The phylocode proposes to throw out Linnaean ranks, too. If there is a worse idea than one-word names for species, this is it. Linnaean ranks, because they are hierarchic, are preadapted to communicate phylogenetic relationships that are hierarchic, too. Rather than pointing to a branching diagram depicting groupings of related species, it is useful to have names for groups like Aves (birds), Compositae (the daisy family) and Arthropoda (insects and relatives). Lacking in humility and common sense, phylocode advocates propose to enshrine current ideas about phylogenetic relationships in immutable names. If you are searching for an antonym for science, you could do no better than phylocode. What if taxonomists before Darwin had been as arrogant? We would have been left with names at odds with the reality of evolutionary history. To believe that our present ideas about relationships are entirely accurate, that they will never be improved upon or replaced in the future, is the definition of chutzpah. Only if systematics ceases to be a science, testing and correcting its ideas, will such inflexible names be preferable. Scientific names for higher taxa in the Linnaean system act as shorthand notations for hypotheses about relationships. These unique identifiers, connected to descriptions, function superbly for information retrieval and communication.

Phylogenetic Classification: General Reference System

Considering the seismic impact of Darwin's theory of evolution, it may be surprising to learn that it had very little effect on how taxonomy was done. Taxonomists were among the first to acknowledge the theory and adopt evolutionary language in their publications. Nonetheless, how they went about recognizing, describing and classifying species changed very little. There was a good reason for this. The patterns of similarities and differences that taxonomists studied hadn't changed, only Darwin's *post hoc* explanation for them. They had been evident before Darwin, and they still were.

The *Origin* was very nearly silent on two fundamental aspects of evolution. First, the exact mechanisms of evolution. How did mutations arise? How was information passed from ancestor to descendant? And second, what was the history of evolution? Realizing that evolution had taken place, it was clear that characters and species had an historical sequence. Unrecognized at the time, a race was on to fill these two gaps in Darwin's book. Unfortunately for taxonomy—but good for biology, of course—geneticists got their breakthrough first. Following the rediscovery of Mendel's garden-pea crossings, around 1900, advances came rapidly, including the idea that genes are located on chromosomes, that DNA was connected to gene transfer and, by the 1950s, that its chemical structure was a double helix. With so many advances in half a century, it is easy to understand why attention and funding were directed to this blossoming discipline.

During the same period, taxonomists continued the struggle to make classifications reflect evolutionary relationships. Lacking theories and methods to do so in a rigorous way, there remained an element of subjectivity and lack of rationale for objectively testing classifications. This, in spite of many patterns so obvious that they produced classifications still accepted today.

It may seem counterintuitive that a classification reflecting evolutionary relationships should be independent of assumptions about evolutionary processes. With so many factors affecting evolution, from selection to sea levels, and no way to associate a process or processes with a particular speciation event, systematists were in a tough spot. Spectacular advances in understanding genetically based processes were of little help in the challenge to make historical patterns testable. To predicate relationships and classifications on assumptions about processes would be to render them tautological. Process assumptions would dictate decisions about grouping that, in turn, would only appear to confirm the processes they were based on.

It would not be until 1966, when Hennig's *Phylogenetic Systematics* was translated into English, that taxonomists got their theoretical breakthrough. His ideas were the first shot in a conceptual revolution that fundamentally changed how taxonomists work. Suddenly, species and higher taxa could be made objectively testable. And classifications could synthesize and organize information consistent with phylogeny and, in so doing, take on predictive powers.

By the time Hennig's theories catapulted systematics into the company of rigorous sciences, the ship had sailed. Enthusiasm about genetic mechanisms had sucked all the oxygen out of the room, diverted funding into genetics and, even more damaging to systematics, established a deeply rooted experimental, process-oriented paradigm in biology. Nearly everything was seen through the lens of function, from molecules to ecosystems. The study of history, sidelined for decades, impossible to study from a process perspective and tarred with the brush of former subjectivity, had been effectively frozen out of the mainstream of biological activities and funding.

A population biology frame of reference, based on the idea of a continuum of genetic variation *within* species, now colors the way biologists think about biodiversity at the species level, too. Steeped in this genetics framework, reinforced by biology's bias for process, many simply assume that species are arbitrarily carved out of a continuous genetic landscape. To understand species requires a different frame of reference. Hennig described this as the difference between tokogenetic relationships (i.e., reticulate relationships among individuals within bisexual populations) and phylogenetic relationships (i.e., relationships among species). Evidence for the former involves the frequency of distribution of variable traits within species; for the latter, it involves shared-derived similarities Hennig called synapomorphies. It was this special kind of similarity that cut through the confusion associated with overall similarity and the mixing of variable traits and fixed characters.

Systematics enjoyed a resurgence of popularity in the 1970s as Hennig's ideas were debated and the use of computer algorithms explored. But this revival was soon to be derailed by genetics again. When DNA sequencing became affordable, this new data, and the flood of money behind it, quickly overshadowed taxonomy's progress. With the prospect of finally learning the story of the diversification of life within reach, the emphasis shifted away from understanding characters to finding the shortest tree depicting species relationships. When the tree became the focus, individual characters were valued less than simply having lots of data. And DNA provided it.

Trees, in the absence of deep knowledge of species, don't mean very much. It is as if we have been given the table of contents of a great book of secrets but denied access to the passages within. For anyone interested in knowing the history of life, it was a tragic development. What is most astounding to me is how many biologists are content to live with a mere framework, unbothered by ignorance of the details that make studying relationships rewarding. This weird lack of interest in complex and novel characters will eventually give way to curiosity, but by then, it may be too late.

Only a revival of systematics, a renewed interest in completing Hennig's revolution and a return of focus to individual species and characters, can set the situation right. Our goal can be nothing less than knowledge of every species that exists, every attribute they possess and a classification that accurately reflects their phylogeny. The historical pattern of relationships among species provides an essential conceptual context for biology. Life is so diverse, and the course of evolution influenced by so many factors, that the only thing that all species share in common is history. This is why a phylogenetic classification is, as Hennig suggested, the most useful general reference system for biology. And why systematics deserves to be restored to a place of prominence among the life sciences. As taxonomic knowledge grows, priorities for conservation will become increasingly clear and we can confront the biggest conservation issue not being talked about.

Further Reading

Abidin, D. H. Z., Nor, S. A. M., Lavoue, S., et al. (2021) DNA-based taxonomy of a mangrove-associated community of fishes in Southeast Asia. *Scientific Reports*, 11: 17800. doi:10.1038/s41598-021-97324-1

Ahrens, D., Ahyong, S. T., Ballerio, A., et al. (2021) Is it time to describe new species without diagnoses?—A comment on Sharkey et al. (2021). *Zootaxa*, 5027: 151–159.

Blanke, A., Koch, M., Wipfler, B., Wilde, F., and Bernhard, M. (2014) Head morphology of *Tricholepidion gertschi* indicates monophyletic Zygentoma. *Frontiers in Zoology*, 11: 16. doi:10.1186/1742-9994-11-16

Carpenter, S. R., Mooney, H. A., Agard, J., et al. (2009) Science for managing ecosystem services: Beyond the millennium ecosystem assessment. *Proceeding of the National Academy of Sciences*, 106: 1305–1312.

Casiraghi, M., Galimberti, A., Sandionigi, A., Bruno, A., and Labra, M. (2016) Life with or without names. *Evolutionary Biology*, 43: 582–585.

Eldredge, N. and Stanley, S. M., eds. (1984) *Living Fossils*, Springer Verlag, New York, 291 pp.

Gaffney, E. (1979) An introduction to the logic of phylogeny reconstruction. In *Phylogenetic Analysis and Paleontology* (eds. J. Cracraft and N. Eldredge), Columbia University Press, New York, pp. 79–111.

Keller, R. A., Boyd, R. N., and Wheeler, Q. (2003) The illogical basis of phylogenetic nomenclature. *Botanical Review*, 69: 93–110.

Lucy, M. (2017) The first photograph of Earth taken from space. *Cosmos*, 24 October. https://cosmosmagazine.com/space/the-first-photograph-of-earth-taken-from-space/ (Accessed 10 January 2023).

Marakeby, H., Badr, E., Torkey, H., et al. (2014) A system to automatically classify and name any individual genome-sequenced organism independently of current biological classification and nomenclature. *PLoS ONE*, 9: e89142. doi:10.1371/journal.pone.0089142

Meierotto, S., Sharkey, M. J., Janzen, D. H., et al. (2019) A revolutionary protocol to describe understudied hyperdiverse taxa and overcome the taxonomic impediment. *Deutsche Entomologische Zeitschrift*, 66: 119–145.

NASA (2022) The Earth Science Data and Information Project. Slide Deck from NASA Airborne and Field Data Workshop, March 29–30. www.earthdata.nasa.gov/s3fs-public/2022-05/2022-03ADMGWorkshop_ESDISFLindsey.pdf (Accessed 10 January 2023).

NCBI (2022) NCBI Insights: GenBank Release 252.0. 19 October. https://ncbiinsights.ncbi.nlm.nih.gov/2022/10/19/announcing-genbank-release-252-0/ (Accessed 10 January 2023).

Nixon, K. C. (2003) The PhyloCode is fatally flawed, and the "Linnaean" system can easily be fixed. *Botanical Review*, 69: 111–120.

Pennisi, E. (2022) Microbiologists propose new DNA-based naming system for microbes. *Science*, 19 September. doi:10.1126/science.ade9610

Pinker, S. (1994) *The Language Instinct: How the Mind Creates Language*, William Morrow & Co., New York, 483 pp.

Pires, M. M., Grech, M. G., Stenert, C., et al. (2021) Does taxonomic and numerical resolution affect the assessment of invertebrate community structure in New World freshwater wetlands? *Ecological Indicators*, 125: 107437. doi:10.1016/j.ecolind.2021.107437

Pons, J., Barraclough, T. G., Gomez-Zurita, J., et al. (2006) Sequence-based species delimitation for the DNA taxonomy of undescribed insects. *Systematic Biology*, 55: 595–609.

Sagan, C. (1977) *The Dragons of Eden*, Random House, New York, 263 pp.

Scoble, M. (2008) Magister. In *Letters to Linnaeus* (eds. S. Knapp and Q. Wheeler), Linnean Society of London, London, p. 236.

Sharkey, M. J. Janzen, D. H., Hallwachs, W., et al. (2021) Minimalist revision and description of 403 new species in 11 subfamilies of Costa Rican braconid parasitoid wasps, including host records for 219 species. *ZooKeys*, 1013: 1–665.

Shreeve, J. (2004) Craig Venter's epic voyage to redefine the origin of species. *Wired*, 1 August. www.wired.com/2004/08/venter/ (Accessed 10 January 2023).

Vandebroek, I., Reyes-Garcia, V., de Albuquerque, U. P., Bussmann, R., and Andrea, P. (2011) Local knowledge: Who cares? *Journal of Ethnobiology and Ethnomedicine*, 7, 35. www.ethnobiomed.com/content/7/1/35 (Accessed 10 January 2023).

Wen, J., Ickert-Bond, S. M., Appelhans, M. S., Dorr, L. J., and Funk, V. A. (2015) Collections-based systematics: Opportunities and outlook for 2050. *Journal of Systematics and Evolution*, 9999: 1–12. doi:10.1111/jse.12181

Wheeler, Q., Knapp, S., Stevenson, D. W., et al. (2012) Mapping the biosphere: exploring species to understand the origin, organization and sustainability of biodiversity. *Systematics and Biodiversity*, 10: 1–20.

Williams, D., Schmidt, M. and Wheeler, Q., eds. (2016) *The Future of Phylogenetic Systematics: The Legacy of Willi Hennig*, Cambridge University Press, Cambridge, 488 pp.

Wilson, E. O. (1984) *Biophilia*, Harvard University Press, Cambridge, 176 pp.

Wilson, E. O. (1992) *The Diversity of Life*, Harvard University Press, Cambridge, 424 pp.Wooley, J. C. and Ye, Y. (2009) Metagenomics: Facts and artifacts, and computational challenges. *Journal of Computational Science and Technology*, 25: 71–81.

Wygodzinsky, P. (1961) On a surviving representative of the Lepidotrichidae (Thysanura). *Annals of the Entomological Society of America*, 54: 621–627.

Zamani, A., Pos, D. D., Fric, Z. F., et al. (2022) The future of zoological taxonomy is integrative, not minimalist. *Systematics and Biodiversity*, 20: 1–14.

13 Other than That, Mrs. Lincoln, How Was the Play?

What is the point of conservation? With numerous individuals, institutions, organizations and government agencies committed to conservation, this may seem an unusual time to ask such a question, but we had better do so—and fast. In spite of decades of good intentions and significant accomplishments, we are still missing what may be the most important thing to establish about biodiversity conservation: What is it, exactly, that we wish to achieve?

Listening to some conservationists, it is easy to form the impression that the aim is to save every species. That was never realistic, of course. And nothing is gained by perpetuating the myth that it is possible at a time of rapid extinction. Questions we need to ask include "How can we reduce the number of species lost?" and, no less importantly, "How can we assure that species surviving are diverse?" That a large number of species will go extinct is a foregone conclusion. Which and how many are yet to be determined. The admission that we cannot save all species demands a different approach.

E. O. Wilson's *Half-Earth* vision may be the last best hope for biodiversity on a global scale. With the potential to save 85% of species, it offers what is likely the most impactful and realistically achievable vision for a biologically diverse planet. But to realize an optimal conservation outcome, we must be clear on the end-game. I suggest that the ultimate goal, stated simply and clearly, is *to minimize the number of species lost, and maximize the diversity of species surviving*. To that end, there are some obvious steps, including completing an inventory of species, understanding phylogenetic relationships and mapping geographic distributions.

Can you imagine allowing a cruise ship to set sail or a commercial airliner to take flight without a manifest, a list of all the souls onboard? Of course not. It would be unforgivably irresponsible. In the event of a disaster, we would have no idea how extensive a rescue mission to organize, no way to know if or when everyone was accounted for. Yet we approach the conservation of biodiversity and protection of ecosystems with no idea which or how many species are onboard. A first step for making a serious commitment to conservation is a manifest of earth's passengers, an accurate list of species and their ecological seat assignments. We will then be in a better position to rescue as many as possible, and account for all.

The road to biodiversity conservation must be paved with knowledge, not good intentions. That road leads somewhere else with a much warmer climate. So far,

DOI: 10.4324/9781003389071-14

we have muddled along with platitudes and surprisingly little knowledge. If we want the best outcome of the biodiversity crisis, we must articulate a clear vision of success. The current piecemeal approach, with sometimes conflicting goals, leaves too much to chance. As Yogi Berra said, unless you know where you're going, you might not get there.

Too many conservation practices are feel-good activities. We say the right things and engage in symbolic tree-hugging, when we ought to be creating scientific knowledge with which to save priority species with intentionality. It is no longer good enough to focus on a few charismatic species when the whole of biodiversity is threatened. Conservation cannot achieve its full potential so long as it is driven more by emotion than facts, or when the enabling mission of systematics is ignored. We must know what exists if we are to responsibly impact what continues to be.

Some of my colleagues are fatalistic, believing that it is too late to inventory all species, much less save the majority of them. With rampant habitat destruction and extinction out of control, they are prepared to surrender without a fight. If continuing on our present path were our only option, discovering no more than 20,000 species each year, I would be forced to agree that learning enough about species to guide us through this crisis is hopeless. But we have the option to rebuild taxonomy, shedding the prejudices and assumptions that have held it back for years.

Books like Wilson's *Half-Earth* and Reid and Lovejoy's *Ever Green* should bolster our spirits and remind us that sparing millions of species and earth's great ecosystems are prizes worth fighting for. Beyond conservation, exploring the diversity and history of life on this planet is worth fighting for, too. For these and other reasons, we must waste no time undertaking a campaign to inventory all species—and their attributes. The aims of conservation and systematics can inform one another as they go. There is no reason to delay the protection of biodiversity hotspots, the creation of corridors between fragments of valued habitat, or ironing out international agreements to remove barriers to scientific collecting. As knowledge of species grows, we will become better and better at prioritizing areas to be set aside for biodiversity. And as natural places are protected for biodiversity, we will be given heart to continue.

Conservation intends to save species, but which ones and how many? For decades, conservation rhetoric has avoided these questions, allowing the public to simply assume we intend to save them all. For an NGO to admit that the plan is not all sunshine and roses could threaten fund raising. But setting expectations higher than what is possible is a dangerous strategy. Better to share realistic goals and the urgency attached to them.

How the U.S. Endangered Species Act has been portrayed in the media has contributed to the myth that we are working to save all species. A perception exists that the recognition of a threatened species triggers a response from the government. Of course, such was never the case. Very few biologically threatened species are ever officially listed as endangered, much less granted a rescue plan. In some environmental circles, it is heresy to state the obvious, that all species cannot and will not be saved. But this is reality. Misleading the public is a risky strategy with the potential to backfire in disastrous ways. Of all things required of the conservation community at this time, credibility ranks at the top.

Society deserves the hard truth. Decisions we will make in the near future carry life and death consequences. If we take the wrong actions, or fail to act, we will seal the fates of hundreds of thousands, perhaps millions, of species. Talk about an inconvenient truth. Denial of this worrisome reality does not discharge our ethical responsibility to do what we can, in the best way we can. Given basic knowledge of species, we gain the power to both mitigate the number of species lost and, no less importantly, influence which species survive. In an ideal world, all species would be given equal priority. It is difficult to think of an ethical justification for valuing the life of one species over another. For practical and selfish reasons, we look favorably on species useful to us and unfavorably on pests and agents of disease, but the ethics of extinction deserve considerations beyond impacts on us. Our ignorance of millions of species in no way absolves us of responsibility for what is happening to them. And given our role in creating the extinction crisis, we should not be able to duck uncomfortable decisions that must now be made. We have created a *Titanic* problem: biodiversity is sinking fast and there are too few lifeboats. It is time to decide who goes to the front of the line and who goes down with the ship.

No right-minded person would wish for this Sophie's choice, to play God and decide which species live or die. Burying our heads in the sand, avoiding the decision, however, amounts to turning this crisis into a toss of the dice. Wilson's proposal, being put into action as the Half-Earth Project, represents the most practical biodiversity conservation strategy on a planetary scale. With a taxonomic renaissance and species inventory, we can create the knowledge with which to select the best number and combination of places to implement his vision to its greatest effect.

The focus used to be on saving charismatic species, like whales, polar bears and sequoias. I am not above using such poster-child species to tug at heartstrings, if it helps. But this doesn't give the public much credit. I believe that people, informed of what is happening, would get behind a meaningful rather than symbolic response. Conservation biologists have wisely shifted tactics away from saving individual species to protecting the habitats where they live. This is far more likely to succeed in the long run, and it sweeps up lots of species in the process. Zoological parks and botanical gardens create an opportunity for the masses to marvel at species they might never see in the wild, but species "saved" from extinction in artificial environments are more like critically ill patients on life support than conserved species. Place-based conservation is the way to go, but which, how many and what sized places? Ecologists, population biologists and others play critical roles in answering these questions, but so too do systematists.

Like a baby step, this geographic basis for conservation has moved us incrementally closer to appropriate biodiversity conservation policies. While professing to care about all species, we continue to emphasize places where cute and cuddly species live. If we care about majestic tigers, then we must save their home. The theory, of course, is that all other species living there are saved in the bargain.

But there is insufficient evidence to conclude that celebrity species zip codes are enough to conserve all the biodiversity we should want to save.

Conservationists are charged only with saving living species. Systematists have a broader mandate. Their purview includes preserving specimens and evidence of all species, alive or dead, as a window into their attributes and origins. Of course, I prefer that every species continue living. But the grim reality is that millions of species are almost certain to disappear before the biodiversity crisis is over in spite of our best efforts. And there is much to be gained for science and society by preserving specimens and observations of every species, regardless of their fate. Natural history museums are the best insurance against a ghastly loss of knowledge, further compounding the tragedy of mass extinction. Only taxonomists have the burden of assuring that museums represent the full diversity of life.

I recall an exchange with a conservationist following a lecture that I delivered on the biodiversity crisis and systematics. Something I said raised the touchy question of collecting specimens of species on the brink of extinction. To me, the only thing worse than losing a species is having no evidence that it ever existed. Someone in the audience asked, "Are you saying that, if you knowingly saw the last living individual of a beetle species, you would collect it?" My response was, much to his horror, "Absolutely!" Unless it is a gravid or parthenogenetic female, the very last individual is not going to reproduce. Allowing it to crawl off, die and rot away under a bush is in no one's interest, not even the beetle. On the other hand, having a well-preserved specimen, even a single one, contributes importantly to science and pays respect to the species. In the real world, commonsense measures could be taken since we would have no way of knowing if the specimen were truly the last of its kind. It could be kept under surveillance, for example. Or, assuming both a male and female existed, captive breeding attempted. I am not suggesting that we hunt down and kill threatened species, but the supposed virtue in allowing a species to go extinct unknown, without a trace, is a false one.

This pragmatic view sounds cold-hearted and goes against the grain of the tree-hugger *ethos*. The magnitude of the crisis we face is so great that we cannot allow sentimentality to impose avoidable ignorance. It is only with facts, knowledge and yes, museum specimens, that we can effectively respond to the extinction crisis and achieve the best conservation and scientific outcomes. We either make symbolic stands that momentarily make us feel virtuous or take informed positions with the potential to achieve higher conservation goals.

The position of many experimental biologists is no less callous in its own way. More than once, I have heard molecular geneticists and ecologists say we should not waste time and money collecting and naming species that will soon be extinct. After all, they selfishly reason, they will not be available for experimental study, so why bother? That scientists could openly argue against knowledge reveals the ugly myopia of experimentalism carried to extremes. A naïve, uninformed person might be excused for espousing this view. But from a professional scientist, it is a symptom of a dangerous lack of curiosity and respect for knowledge. Species have become, for them, no more than lab rats. And they have no use for dead ones. Presumably, fields like paleontology, and a good deal of anthropology, are of questionable value to them as well. Think how impoverished our lives would be without knowledge of the fossil record or early hominids!

The habitat model of conservation is not as simple as it sounds. The areas saved must be of sufficient size to support enough genetic diversity to avoid inbreeding, and, if they are separate parcels, there must exist habitat bridges by which gene flow can occur. Unaware of the population size requirements of each species in an ecosystem, it is reasonable to gauge minimum areas based on the needs of species, like top predators, thought to have the largest spatial requirements. This is clearly the best guess for tract sufficiency, but without baseline data about many other species, we can't be certain.

Conservation should be more than a numbers game, because species are not "created" equally. Consider two hypothetical islands, Isla Alpha and Isla Beta, each inhabited by ten species. Isla Alpha has one species each of palm tree, fern, turtle, mouse, bird, fly, beetle, earth worm, mite and mushroom. Isla Beta has two species of palm trees and eight species of palm weevils. In numbers of species, the two islands are precisely equal. But in *phylogenetic* diversity, they are quite different. Saving Isla Alpha conserves members of ten major lineages, while saving Isla Beta conserves only two. Conservation should be informed by phylogenetic diversity, as well as numbers, and may include visualizing phylogenetic diversity in relation to GIS as described by Mishler. Kling *et al.* go farther, arguing that conservation targets take account of lineage diversification, character divergence and survival time as facets of phylogenetic diversity. These and similar efforts enrich our understanding of biodiversity and make conservation goals more precisely defined.

As a taxonomist who is fascinated by species, it pains me to say so, yet conserving some species should be a higher priority than saving others. The tropical beetle genus *Agra* includes about 2,000 arboreal species. The tree family Ginkgoaceae, once diverse, includes a single surviving species, *Ginkgo biloba*. An assessment of the situation would suggest that losing the last living ginkgo is, in some real sense, a greater loss to science and the diversity of the biosphere than the loss of one among thousands of species of *Agra*. Saying such things makes you as welcome among conservationists as an undertaker in a nursing home. But as we develop a planetary conservation master plan, avoiding such choices is irresponsible.

To the casual observer, there are a couple of caricatures of conservation, neither fair to those working diligently to save biodiversity. In one distortion, conservationists care only for pretty species and pristine places, paying little attention to much of the diversity of life. In the other, they have a kneejerk response to news of any species being threatened, immediately bringing development to a standstill. There are compelling reasons to assure that species remain as diverse as possible, but we are way past the point at which it is realistic to stop all land development that encroaches upon biologically threatened species. If we knew all species, there would be very few places on the globe where at least a few threatened ones did not occur. The magnitude of the crisis we face is too great to justify continuing this charade.

It is imperative that we clarify the overarching goal of conservation as the guiding principle for a master plan. What does a successful end-game look like?

If we wish to conserve as much diversity of life as possible, and we are not content with a simple head count, then it is even more important that we learn what species there are, what attributes they have and how they are related to one another. That, of course, is the bread and butter of systematics. And for credible, testable knowledge, there are no short cuts.

There is understandable fear in the conservation community that admitting that all species cannot be saved would put the entire enterprise on a slippery slope. If it is acceptable to write off one species among 10 million, why not a thousand? And if a thousand, who will miss a million or two if we can have more shopping malls and oil palm plantations? Yet in science, as in all things, honesty is the best policy. I think that it is time that we stop pretending that we can or will save every endangered species and make the hard decisions called for to assure that conservation produces the results we want and that are best for the biosphere.

Systematics part in minimizing extinctions and maximizing diversity includes determining what species exist and where they live, how they are phylogenetically related and how to identify them so that we can assess their relative contribution to diversity and keep track of their status. This gives conservation biology the basic knowledge with which to develop informed, measurable goals. Systematics and conservation biology are natural partners, each with important, yet distinct, responsibilities. Conservation biology is a complex science, drawing knowledge from systematics, ecology, population genetics, geography, economics, politics and sociology to achieve its goals. Systematists have a different role to play exploring, describing, naming, classifying and mapping species. In making species known, taxonomists lay the foundation for conservation and increase the chances that the Half-Earth Project can succeed. Only with knowledge of species, relationships and distributions is it possible to create a calculus of conservation and arrive at the optimal combination of places to conserve the greatest diversity of species.

Morality, Ethics and Conservation

Every species, including our own, will eventually go extinct but that is no reason to hurry the inevitable along. Human actions have poured accelerant on extinction. As a result, rates of species loss are alarmingly faster than in prehistory. Playing a major role in this mess, it falls upon humankind to examine the morality of its continuing actions and its responsibility to mitigate this disaster by reducing its impacts.

It can be argued that saving biodiversity is the right thing to do for many reasons, not least of which is the fact that the biosphere and its ecosystems make possible our existence. We sell ourselves short, however, when we justify conservation by emphasizing ecological services. There are many benefits of keeping a species-rich world. A biologically diverse planet is filled with beauty and wonder, the mystique of the unknown and boundless possibilities. Our lives are elevated in unexpected ways by all kinds of species with nothing of obvious value to offer us.

In *Cosmos*, Carl Sagan said that "Compared to a star, we are like mayflies, fleeting ephemeral creatures who live out their whole lives in the course of a single day." In the extreme, there are species of mayfly that live only minutes as adults, never knowing the pleasure of a single meal, prompting us to question what we are achieving with the comparative eternity of a human life span. The dazzling diversity of color and form among the 75 species of tulips, and more than a thousand hybrids and cultivars, is prized around the world for their inspiring beauty. And living in the company of animals is a reminder that, in important respects, *Homo sapiens* differs from them only by degree … and not always to our credit. In Edinburgh, Scotland, there is a statue and plaque honoring a Skye Terrier named Bobby who, for 14 years after his master's death, refused to leave his grave. The inscription reads "Greyfriars Bobby, died 14th January 1872, aged 16 years, let his loyalty and devotion be a lesson to us all."

As Georgina Mace said

> Taxonomy and conservation go hand in hand. We cannot necessarily expect to conserve organisms that we cannot identify, and our attempts to understand the consequences of environmental change and degradation are compromised fatally if we cannot recognize and describe the interacting components of natural ecosystems.

As an inventory of species progresses, we will confront the hard facts of extinction and no longer be able to deny what is happening to biodiversity. As we contemplate what is being lost, we owe it to ourselves to recall all the incredible ways in which our lives are enriched by the existence of other species, physically, economically, intellectually and dare I say spiritually. Facing up to the unthinkable, determining which species live or die, we will discover how well our moral compass is working and be challenged to clarify for ourselves the principles we wish to live by.

What rules do we aspire to follow in our interactions with nature and attempts to balance the biosphere's needs against our own? In the past, we have rationalized pillaging biodiversity to increase our short-term comfort or wealth, but the time has come to recalibrate our relationship to the natural world. Rather than finding ourselves, in the not-too-distant future, eking out a hard-scrabble existence on a barren planet, living in deep regret, it makes sense to be introspective today. Let's step back and examine the ethical dimensions of our relationship to other species, our awesome power to reshape the biosphere and our selfish nature that has so far kept us from taking the long view. We owe it to ourselves to be more thoughtful. And to be resolute in our mission to learn the millions of species that our acts are impacting. It is not enough to have data, information and knowledge. We need wisdom, too, and an ethical code of conduct that is well considered and altruistic. When it comes to the biosphere, we broke it and now own a biodiversity crisis. There is no way to undo much of the damage we have caused, but we can at least respond from here forward in ways that we can be proud of.

Further Reading

Bailey, L. H. (1915) *The Holy Earth*, C. Scribner's Sons, New York, 171 pp.

Cardinale, B., Primack, R. B., Murdoch, J. D., and Murdoch, J. (2019) *Conservation Biology*, Oxford University Press, Oxford, 672 pp.

Erwin, T. L. (1991) An evolutionary basis for conservation strategies. *Science*, 253: 750–752.

Forey, P. L. and C. J. Humphries (1994) *Systematics and Conservation Evaluation*, Clarendon Press, Oxford, 466 pp.

Hawksworth, D. L., ed. (1995) *Biodiversity: Measurement and Estimation*, Chapman and Hall, London, 140 pp.

Hunter, M. L. and Gibbs, J. P. (2007) *Fundamentals of Conservation Biology*. 3rd ed., Wiley-Blackwell, Malden, 497 pp.

Idzikowski, L., ed. (2020) *Biodiversity and Conservation*, Greenhaven, New York. 176 pp.

de Klemm, C. (1993) *Biological Diversity Conservation and the Law*, IUCN, Gland, Switzerland, 292 pp.

Kling, M. M., Mishler, B. D., Thornhill, A. H., et al. (2018) Facets of phylodiversity: Evolutionary diversification, divergence and survival as conservation targets. *Philosophical Transactions of the Royal Society* B, 374: 20170397. doi:10.1098/rstb.2017.0397

Mace, G. M. (2004) The role of taxonomy in species conservation. *Philosophical Transactions of the Royal Society of London, B*, 359: 711–719.

Maxted, N., Ford-Lloyd, B. V., and Hawkes, J. G., eds. (1997) *Plant Genetic Conservation: The In Situ Approach*. Kluwer Academic Publishers, Dordrecht, 446 pp.

Mishler, B. D. (2023). Spatial phylogenetics. *Journal of Biogeography*, in press.

Ninan, K. N., ed. (2009) *Conserving and Valuing Ecosystem Services and Biodiversity: Economic, Institutional and Social Challenges*, Earthscan, London, 402 pp.

Nixon, K. C. and Wheeler, Q. D. (1992) Extinction and the origin of species. In *Extinction and Phylogeny* (eds. M. J. Novacek and Q. Wheeler), Columbia University Press, New York, pp. 119–143.

Prathapan, K. D., Pethiyagoda, R., Bawa, K. S., et al. (2018) When the cure kills—CBD limits biodiversity research. *Science*, 360: 1405–1406.

Reid, J. W. and Lovejoy, T. E. (2022) *Ever Green: Saving Big Forests to Save the Planet*, W. W. Norton, New York, 320 pp.

Ross, J. and Adkins, R. (2018) *Biodiversity and Environmental Conservation*, ED-Tech, Waltham Abbey, 348 pp.

Sagan, C. (1980) *Cosmos*, Random House, New York, 365 pp.

Thomson, S. A., Pyle, R. L., Ahyong, S. T., et al. (2018) Taxonomy based on science is necessary for global conservation. *PLoS Biology*, 16: e2005075. doi:10.1371/journal.pbio.2005075

Wheeler, Q. D. (1995) Systematics and biodiversity: Policies at higher levels. *BioScience*, 45: s21–s28.

Wheeler, Q. D. and Novacek, M. J. (1992) *Extinction and Phylogeny*, Columbia University Press, New York, 253 pp.

Part II

A Crisis of Crises

As we face environmental challenges unprecedented in kind, scale, number and complexity, we are hampered by four separate, mutually reinforcing, crises: a biodiversity crisis, with the rate of species extinction accelerated by a growing population, the degradation and conversion of natural ecosystems, and climate change; a systematics crisis, with critical losses of the expertise, funding and institutional leadership needed to complete an inventory of life on earth, before it is too late; a crisis of awareness and ethics, magnified by a widening gap between humankind and the natural world; and a dearth of options for adapting to environmental change and creating a sustainable future. Our best course of action is to confront these crises with clear goals: to minimize the number and diversity of species lost, revitalize systematics, organize a planetary-scale inventory, encourage up close and personal interactions with nature and lay the ground work for an evolutionary economy that adapts to whatever comes, and never ceases to innovate in a quest to maintain both human prosperity and as much of the natural world as possible.

DOI: 10.4324/9781003389071-15

14 Extinction

"911, what's your problem?"
"The Library of Congress is on fire!"
"How many books have been lost?"
"Uh … well … I don't know exactly … Please, please, just send the firefighters!"
"How can we be sure this is serious enough to commit resources? Books are lost all the time. They get shelved in the wrong place, don't get returned, they're damaged in accidents. Call me back when you know how many books are in imminent danger. Have a nice day."

There are skeptics who question the seriousness of the extinction crisis in the absence of hard numbers and based on what they have personally witnessed or, rather, not witnessed. Lulled into complacency by the fact that species have existed in abundance throughout human history, they want tangible evidence of a crisis. They want to see bodies with toe tags, not hysterical rhetoric about massive numbers of unseen, unnamed species that have not yet been shown to exist, much less being or becoming extinct. For the data wonk, they have a point. After all, little more than a thousand species extinctions have been well documented in historic time. A thousand extinct species is unfortunate to be sure, but in a world with at least 10 million species, hardly a hair-on-fire crisis.

Elizabeth Kolbert eloquently described our situation: "Right now, in the amazing moment that counts as the present, we are deciding, without quite meaning to, which evolutionary pathways will remain open and which will forever be closed." Since humans have paid attention to such things, we have gathered firm evidence of the extinction of only about 500 plants and 700 vertebrate animals. Yet I know no taxonomist, ecologist or conservation biologist who believes this reflects more than a minute fraction of the actual number of extinctions. One frequently cited estimate of the current rate is around 20,000 extinctions per year. And Cowie et al. suggest that 10–15% of the 2 million named species are already gone.

Beyond numbers, taxonomists have made two interrelated observations that add up to bad news for species. First, millions of species are narrowly distributed geographically, ecologically or both. Second, species they depend upon or the places they live—such as one or a few host plants, or isolated islands or mountaintops—are

DOI: 10.4324/9781003389071-16

being lost, degraded in quality or hopelessly modified. It follows that as unique habitats go, so go species living there and nowhere else.

The rate of extinction, as you might expect, varies greatly among taxonomic groups and from place to place. Some groups are little affected so far. Others are in grave danger. The worst extinction rates vary from a few hundred times the "normal" background rate—that is, the *tempo* of species disappearance observed in the fossil record—to thousands of times that rate. We should not have to wait until millions of species are gone, until rock solid evidence of their extermination exists, before concluding that there is a crisis and responding accordingly. Looking at the situation overall, Peter Raven and Scott Miller come to a sobering conclusion:

> Unless we control the underlying causes, including overdevelopment and climate change, we are in danger of losing 80% or more of the world's species, the proportion lost 66 million years ago when the dinosaurs became extinct and many of the plants and animals known today began their ascent. *We have clearly entered the world's sixth major extinction event.*
>
> (italics mine)

Besides, with respect to an inventory, rates of extinction really don't matter. If, as scientists believe, millions of species are lost over the next century, then an inventory will have preserved evidence with which scientific exploration of biodiversity, evolutionary history and the primal organization of the biosphere can continue. If by some miracle, on the other hand, they are wrong, and most species go on living, then we shall have made them known and accessible for a host of purposes, from enjoyment to biomimicry and development of new natural resources. In life, we are presented with few such no-lose situations. The return on investment, paid in dividends of knowledge, makes a species inventory a savvy move whatever may come.

I am convinced by anecdotal accounts, personal experience, indirect evidence and common sense that a huge number of species have gone extinct already and that many more will soon follow, some of them today. Direct proof in many cases may not exist, but indications of this catastrophe are everywhere. Delaying action until we have the kind of convincing documentation that exists for a handful of species, like the passenger pigeon, will sign the death warrants for millions of others. As this quote from Ehrlich and Ehrlich suggests, simply seeing what is happening, regardless of its precise degree of advance, ought to be cause enough for concern:

> As you walk from the terminal toward your airline, you notice a man on a ladder busily prying rivets out of its wing. Somewhat concerned, you saunter over to the river popper and ask just what the hell he's doing.

Philadelphia botanists John and William Bartram discovered a flowering tree along the Altamaha River in the Georgia colony in 1765. Growing the tree from seed, William assigned it to a genus he named for Benjamin Franklin. In spite

Figure 14.1 The Tree of Life, Pruned. This harshly cropped specimen in Greenwich, England, standing against a stormy sky, is a metaphor for the biodiversity crisis. As clouds of extinction gather, species disappear first, then entire branches, until the tree of life is unrecognizable. Photo: the author.

of efforts to find it, *Franklinia alatamaha* was last seen in the wild in 1803. Every known living specimen is descended from Bartram's garden. Other species are positioned to follow suit. Botanists estimate that the survival of one-third of the world's tree species is threatened.

Over the past three centuries, taxonomists have found that many species are restricted in respect to where they live. *Orthohalarachne attenuata* is a mite that lives in the nasal passages of fur seals, sea lions and walruses. There is one incidental record of the mite being removed from the eye of a man who was inconsiderately snorted on by a walrus at Sea World. Such accidental transmissions are not the preference of the recipient or the mite and do not increase places where the mite can live and breed. If these ocean mammals were to disappear tomorrow, so would this mite. *Lycoperdina ferruginea* is a beetle that feeds only on puffballs. And

Monotropa uniflora, the Indian Pipe, is a ghostly, unpigmented flower that, although widespread geographically, can survive only in association with a few species of mycorrhizal fungi in the family Russulaceae, often in association with beech trees. The blind, unpigmented amphipod *Stygobromus albapinus* is recorded only from pools in the Model Cave in Great Basin National Park, Nevada. Even within vast ecosystems, like the Amazon basin, there exist intricate spatial patterns of species distributions, including extraordinarily narrow ecological niches. In general, widespread species are less susceptible to extinction than those with small, localized populations, but none are immune.

To pick one admittedly extreme geographic example, consider the island nation of Madagascar. A few statistics tell the story. Ninety percent of the natural habitats of the island have been destroyed or degraded beyond recognition since humans first showed up about 2,000 years ago. Of Madagascar's species, 83% of plants occur nowhere else on earth. Excluding ferns, this percentage increases to 92%! Such uniqueness of species is not limited to the flora, with equally impressive proportions of endemic animals: 92–100% of terrestrial vertebrates, 52–60% of birds and bats and 86% of macroinvertebrates. With only 10% of the original forest cover intact, imagine the number and diversity of species already lost! The same may be said of Brazil's Atlantic coastal forest. These are but two among a depressingly large number of unique places around the world that have been destroyed to greater or lesser degrees.

From frequent examples of host specificity, habitat restriction and geographic isolation, it follows that when a host species or isolated place is destroyed, the species that live there, and only there, are erased, too. Keeping this in mind, one need only examine satellite images or visit any number of places around the world to see how much destruction has been done or is underway. It is simply impossible to observe what is happening throughout the biosphere and conclude anything but widespread extinction. The fact that taxonomy has not been supported to document species is no excuse for denying extinctions. And, ignorance is no justification for delaying a response to the extinction crisis, beginning with a long overdue inventory of species. If you insist on hard proof, then support taxonomists to go get it. And if you don't require it, support them anyway.

Species do go extinct in the usual course of events, with or without our help. Our idea of a "normal" extinction rate, the average during prehistoric time, is based on the fossil record. That record, however, is skewed toward groups prone to be preserved; groups like reptiles, mammals, marine invertebrates with shells and diatoms. We know that even their record is incomplete, sometimes grossly so. Nonetheless, it's the best evidence we have.

The geologic background extinction rate is often cited as 2 species per 10,000 per 100 years. The rate of mammalian extinction over the last century was about 100 times that rate. Not all species go extinct at the same rate, of course. Amphibians, for example, appear to be under much greater threat than most. Assuming that every species of amphibian currently at or near threatened status proceeds to go extinct (while hoping, of course, that the unthinkable will not come to pass), their rate of extinction this century would be at least 25,000 times the background rate! It is no stretch of credulity to assume that amphibians are not alone, that there are other, less well-studied, groups of plants and animals facing similar, or possibly worse, threats. Waiting until we gather enough data to precisely measure the risks faced by millions of species will almost certainly result in massive losses that might have been avoided. All things considered, waiting for better data is the worst strategy of all.

The vast majority of species are not vertebrate animals, so what of them? Estimates exist for some. Seed-bearing plant species are said to be disappearing at a rate 500 times greater than in prehistory. Numbers for invertebrate animals are particularly inadequate, but we know the same principles apply to them, too. Of butterflies, about 90% have specific host associations as larvae. As host plants go, so go butterflies. Irrefutable evidence exists for only about 70 insect extinctions, but this number is ridiculously low. With so many insects, and so few people studying them, who would notice their absence? One extremely conservative estimate, arrived at indirectly, suggests that up to 1,400 insect species have gone extinct since 1600. But other estimates, and common sense, suggest that the numbers are much higher with insects figuring prominently among the estimated 20,000 annual extinctions. For example, among herbivorous insects in biodiversity hot-spots, geographic areas with exceptionally high numbers of species, it has been estimated that as many as half a million species of monophagous insects (those dependent on single host plants) are already on the path to extinction. If insect species are disappearing in roughly the same proportion as they are represented among species discoveries, it would not be good news for them. Each year about half of newly named species are insects and related arthropods. To this, add an alarmingly rapid decline in insect populations. In areas of intense agriculture, there was a 75% decline in insect abundance in less than 30 years. The decline in insects is so noticeable that there is an entomology meme, dubbed the "windshield anecdote" by John Acorn. Since 2000, it has been noticed that fewer bugs are splattered on windshields. One study in Denmark recorded an 80% decline of insects based on bug splatter. How much more evidence do we need?

As an undergraduate and aspiring entomologist, one of my summer jobs was collecting mosquitoes for the Ohio department of health, so they could be tested for Eastern Equine Encephalitis. When I asked people for permission to collect insects on their property, a typical response was "Yes! Please, get 'em all." This attitude toward insects is a dangerous one. From pollination to decomposition, herbivory and predation, insects are essential to just about every terrestrial and freshwater ecosystem, and certainly to human welfare. Given the paucity of data on extinctions of larger animals and plants, it is not surprising that evidence of the status of insects is thinner. Like extinction writ large, however, the information that exists points unmistakably to an alarming trend.

Any school child can tell you about the plight of the monarch butterfly whose populations have declined by 90% in recent decades. Even before the environmental pressures on insect populations today, there were cases of abundant species disappearing. One of the most celebrated was *Melanoplus spretus*, the Rocky Mountain Locust. Their migratory flights were so dense that they literally darkened the sky, turning day to night. In 1875, a 198,000 square mile mega-swarm was estimated to have contained more than 12 trillion individuals. In spite of this once super-abundant status, no confirmed sightings of a Rocky Mountain Locust have taken place since 1902. It has been suggested that as many as 40% of insect species may be in decline. Oliver Milman's *The Insect Crisis* explains in stark terms what is happening to insects and why we should all be worried. Existential threat doesn't seem strong enough. As E. O. Wilson said, without insects, "the world would return to the state of a billion years ago, composed primarily of bacteria, algae and few very simple multicellular plants."

To be honest, I don't believe any of the numbers associated with rates of extinction—but I do trust what they suggest. Things are dire and worsening. I don't have to see an actual foot race between a tortoise and hare to believe in the moral of the fable. An undetermined, but very large, number of species are going extinct annually, with millions at risk over the coming century. Unless we act immediately, that rate seems destined to accelerate in cascading fashion, making the predicted outcome worse or sooner, possibly both.

For arguments sake, what if there were no imminent threat of extinction? What if species were disappearing at something close to the "normal" rate that existed before humans were around? I would still be writing this book, still arguing for a taxonomic renaissance, species inventory and return of natural history museums to their rightful function as centers of excellence for species exploration. I would have a different narrative, of course. There would be less urgency, perhaps, but the benefits of an inventory would be even greater in some respects. The major difference is that having all species named, described and classified would throw open wide the opportunity to enjoy, and benefit from knowing, millions and millions of living species. The more, and more diverse, species that survive, the more we benefit. Thus, an inventory is fantastically important with or without a mass extinction. Further, an inventory is the single best way to arm ourselves with knowledge to avoid some environmental challenges, and mitigate or adapt to others.

This conclusion bears restating. I reject the argument that because we lack hard numbers for actual extinctions that there is no crisis. Indirect data convinces me, and every biologist I know, of the fact that a mass extinction is well underway. The percentage of land that is being, or has already been, converted from native habitat to agriculture or other human ends, trends in deforestation, ocean acidification, natural resource harvesting, measurable effects of climate change and satellite images over time, among other observations, all tell the same story. As human population increased from a million people 12,000 years ago to about 8 billion today, and 10 billion in another 10–20 years, it should not be surprising that the natural world suffered, shrank and declined in ecological integrity. Peter Raven has estimated that were the entire world's population to live at the level of affluence as we in the U.S., four planets the size of earth would be needed to meet their needs! Clearly, the only hope to save a significant amount of biodiversity is to act decisively, and now. I personally believe Wilson's *Half-Earth* vision is the most promising and practical route proposed to date. But delayed support for taxonomy means compromises in its realization.

Beyond confronting extinction, two great challenges deserve our attention. One is to fully document the diversity of life at and above the species level to create legacy knowledge of evolutionary history, biodiversity and the biosphere. The other is to address the need for sustainable ways to meet human needs; ways in which the natural world need not be destroyed or used up in the process; ways in which most of humanity can avoid living in abject poverty so that a few of us can live in comparative luxury. The only way I see to do so in a short period of time is biomimicry. And the surest way to accelerate biomimicry is to document attributes of species, including those about to become extinct. Something molecular data cannot do.

Even accepting that we are in full-blown crisis mode, all is not gloom and doom. It turns out that fewer species go extinct than one might have expected when a given area of habitat is lost. Like the battleground of evolutionary history, strewn with extinct species, the current crisis will take a heavy toll on species that cannot adapt to rapid environmental change, small populations of limited genetic resiliency and species forced to migrate, finding themselves suddenly confronted by new predators and competitors.

Almost every taxonomist can tell you stories about newly discovered species on the brink of extinction. One botanical colleague collected several new tree species from Madagascar and, before the ink of her publication was dry, they were apparently gone in the wild. Joseph McHugh and I named a new species of beetle, *Dasycerus maculatus*, from the high elevation spruce-fir forest in the southern Appalachian Mountains, where it feeds on fungal hyphae on the underside of rotting logs. An adelgid pest introduced from Europe has killed off more than 90% of the fir trees since the 1960s. Some regrowth seems to have resistance to the pest, but the status of other species, like our beetle, remains tenuous.

It is not hyperbole to say that we have entered a period of extinction unlike any since the Cretaceous-Tertiary boundary event 65 million years ago. Because we have neglected taxonomy, museum collections and the inventory of species begun by Linnaeus, we confront this environmental catastrophe blinded by ignorance. We have so little knowledge that we cannot say what species exist to begin with, much less which or how many go extinct in any given year, or where our conservation efforts can do the greatest good. We are guessing when we could, and should, be making fact-driven decisions. We are rolling the dice when taxonomic knowledge could make choices more certain. We have neglected systematics far too long and, if we are wise, we will make the revival of species exploration a centerpiece of 21st century science removing obstacles in its way—including a disproportionate emphasis on molecules.

Further Reading

Acorn, J. (2016) The windshield anecdote. *American Entomologist*, 62: 262–264.

Arndt, N. and Pinti, D. L. (2011) Mass extinctions. In *Encyclopedia of Astrobiology* (ed. M. Gargaud), Springer, Berlin, doi:10.1007/978-3-642-27833-4_547-3

Boehm, M. M. A. and Cronk, Q. C. B. (2021) Dark extinction: The problem of unknown historical extinctions. *Biology Letters*, 17: 20210007. doi:10.1098/rsbl.2021.0007

Ceballos, G., Ehrlich, P. R., Barnosky, A. D., et al. (2015) Accelerated modern human-induced species losses: Entering the sixth mass extinction. *Science Advances*, 1: e1400253. doi:10.1126/sciadv.1400253

Cooke, S. B., Davalos, L. M., Mychajliw, A. M., Turvey, S. T., and Upham, N. S. (2017) Anthropogenic extinction dominates Holocene declines of West Indian mammals. *Annual Review of Ecology, Evolution, and Systematics*, 48: 301–327.

Cowie, R. H., Bouchet, P., and Fontaine, B. (2022) The sixth mass extinction: Fact, fiction or speculation? *Biological Reviews*, 97: 640–663.

Dunn, R. R. (2005) Modern insect extinctions, the neglected majority. *Conservation Biology*, 19: 1030–1036.

Ehrlich, P. R. and Ehrlich, A. H. (1981) *Extinction: The Causes and Consequences of the Disappearance of Species*. Random House, New York, 305 pp.

Einhorn, C. (2020) Wildlife collapse from climate change is predicted to hit suddenly and sooner. *The New York Times*, 15 April.

Farrell, B. D. and Erwin, T. L. (1988) Leaf-beetle community structure in an Amazonian rainforest canopy. In *Biology of Chrysomelidae* (eds. P. Jolivet, E. Petitpierre, and T. H. Hsiao) Kluwer, Dordrecht, pp. 73–90.

Fonseca, C. R. (2009) The silent mass extinction of insect herbivores in biodiversity hotspots. *Conservation Biology*, 23: 1507–1515.

Goodman, S. M. and Benstead, J. P. (2005) Updated estimates of biotic diversity and endemism for Madagascar. *Oryx*, 39: 73–77.Hallmann, C. A., Sorg, M., Jongejans, E. et al. (2017) More than 75 percent decline over 27 years in total flying insect biomass in protected areas. *PLoS ONE*, 12: e0185809. doi:10.1371/journal.pone.0185809

Hannah, L., ed. (2012) *Saving a Million Species: Extinction Risk from Climate Change*. Island Press, Washington, 416 pp.

He, F. and Hubbell, S. P. (2011) Species-area relationships always overestimate extinction rates from habitat loss. *Nature*, 473: 368–371.

Jarvis, B. (2018) The insect apocalypse is here. *New York Times Magazine*, 27 November.

Kahn, J. (2018) Should some species be allowed to die out? *The New York Times Magazine*, 13 March.

Kolbert, E. (2014) *The Sixth Extinction*, Henry Holt, New York, 336 pp.

Lawton, J. H. and May, R. M., eds. (1995) *Extinction Rates*, Oxford University Press, Oxford, 233 pp.

Ledford, H. (2019) Global plant extinctions mapped. *Nature*, 570: 148–149.

Lockwood, J. A. (2004) *Locust: The Devastating Rise and Mysterious Disappearance of the Insect that Shaped the American Frontier*, Basic Books, New York, 294 pp.

Louca, S., Shih, P. M., Pennell, M. W., et al. (2018) Bacterial diversification through geological time. *Nature Ecology & Evolution*, 2: 1458–1467.

Lyell, C. (1832) Changes caused by the progress of human population, In *Principles of Geology: Being an Attempt to Explain the Former Changes of the Earth's Surface, by Reference to Causes Now in Operation*, vol. 2, John Murray, London, 332 pp.

MacLeod, N. (2013) *The Great Extinctions: What Causes Them & How They Shape Life*, Firefly Books, Buffalo, 208 pp.

McCallum, M. L. (2007) Amphibian decline or extinction? Current declines dwarf background extinction rate. *Journal of Herpetology*, 41: 483–491.

McDermott, A. (2020) To understand the plight of insects, entomologists look to the past. *Proceedings of the National Academy of Sciences*, 118: e2018499117. doi:10.1073/pnas.2018499117

Milman, O. (2022) *The Insect Crisis: The Fall of the Tiny Empires that Run the World*, W. W. Norton, New York, 260 pp.

Møller, A. P. (2019) Parallel declines in abundance of insects and insectivorous birds in Denmark over 22 years. *Ecology and Evolution*, 9: 6581–6587. doi:10.1002/ece3.5236

Myers, N., Mittermeier, R. A., Mittermeier, C. G., da Fonseca, G. A. B., and Kent, J. (2000) Biodiversity hotspots for conservation priorities. *Nature*, 403: 853–858.

New York Times (2018) The insect apocalypse is here. *The New York Times*, 27 November.

New York Times (2019) Life as we know it: Plant and animal species are disappearing faster than at any time in recorded history. We know who is to blame. *The New York Times*, 11 May.

Pakaluk, J. (1984) Natural history and evolution of *Lycoperdina ferruginea* (Coleoptera: Endomychidae) with descriptions of immature stages. *Proceedings of the Entomological Society of Washington*, 86: 312–325.

Payne, J. L. and Finnegan, S. (2007) The effect of geographic range on extinction risk during background and mass extinction. *Proceedings of the National Academy of Sciences*, 104: 10506–10511. doi:10.1073/pnas.0701257104

Pimm, S. L. and Raven, P. H. (2019) The state of the world's biodiversity. In *Biological Extinction: New Perspectives* (eds. P. Dasgupta, P. H. Raven, and A. L. McIvor), Cambridge University Press, Cambridge, pp. 80–112.

Piper, R. (2009) *Extinct Animals: An Encyclopedia of Species that have Disappeared*. Greenwood Press, Westport, 204 pp.

Randall, L. (2015) *Dark Matter and the Dinosaurs*, HarperCollins, New York, 417 pp.

Raven, P. H. and Miller, S. E. (2020) Here today, gone tomorrow. *Science*, 370: 149.

Raven, P. H. and Wagner, D. L. (2021) Agricultural intensification and climate change are rapidly decreasing insect biodiversity. *Proceedings of the National Academy of Sciences*, 118: e2002548117. doi:10.1073/pnas.2002548117

Regnier, C., Guillaume, A., Lambert, A., et al. (2015) Mass extinction in poorly known taxa. *Proceedings of the National Academy of Sciences*, 112: 7761–7766.

Sánchez-Bayo, F. and Wyckhuys, K. A. G. (2019) Worldwide decline of the entomofauna: A review of its drivers. *Biological Conservation*, 232: 8–27.

Sartore, J. (2017) *The Photo Ark: One Man's Quest to Document the World's Animals*, National Geographic, Washington, 400 pp.

Stork, N. E. (2010) Re-assessing current extinction rates. *Biodiversity Conservation*, 19: 357–371.

Stork, N. E., McBroom, J., Gely, C., and Hamilton, A. J. (2015) New approaches narrow global species estimates for beetles, insects, and terrestrial arthropods. *Proceedings of the National Academy of Sciences*, 112: 7519–7523.

Taylor, S. J. and Holsinger, J. R. (2011) A new species of the subterranean amphipod crustacean genus *Stygobromus* (Crangonyctidae) from a cave in Nevada, USA. *Subterranean Biology*, 8: 39–47.

Wackermagel, M., Lin, D., Evans, M., Hanscom, L., and Raven, P. (2019) Defying the footprint of oracle: Implications of country resource trends. *Sustainability*, 11: 2164. doi:10.3390/su11072164

Wagner, D. L., Grames, E. M., Forister, M. L., Berenbaum, M. R., and Stopak, D. (2021) Insect decline in the Anthropocene: Death by a thousand cuts. *Proceedings of the National Academy of Sciences*, 118: e2023989118. doi:10.1073/pnas.2002549117

Wagner, D. L., Rox, R., Salcido, D. M., and Dyer, L. A. (2021) A window to the world of global insect declines: Moth biodiversity trends are complex and heterogeneous. *Proceedings of the National Academy of Sciences*, 118: e2002549117. doi:10.1073/pnas.2002549117

Webb, J. P. Jr., Furman, D. P., and Wang, S. (1985) A unique case of human ophthalmic acariasis caused by *Orthohalarachne attenuata* (Banks, 1910) (Acari: Halarachnidae). *Journal of Parasitology*, 71: 388–389.

15 Systematics under Siege

How can we end the siege on descriptive systematics? Systematists have been hunkered down in their bunkers too long. It's time to muster the troops on the battlefield. Because there is nothing fundamentally lacking in the theories and methods of systematics, there are few internal changes required to strengthen its defense against external forces suppressing its mission. What is most needed is simple: refute false narratives about the discipline, reembrace the mission of systematics, restore funding and positions, allow taxonomists to do taxonomy and get out of the way. Because of advances in technology, there are few tasks of the taxonomist that cannot be done faster and better than before. Even with a historically unprecedented extinction rate, the greatest challenge for systematics is sociological. Scientists and administrators with influence over hiring and funding priorities are, as a rule, not malicious. Unaware of their misconceptions about taxonomy, they are, however, vulnerable to bias and groupthink.

Making matters worse, the prejudice against taxonomy is reinforced by ideas that have deep roots. We have mentioned experimental biologists who have a rigid idea about what good science looks like. And that DNA is preferred, not because it tells us more about species or is inherently more interesting or reliable than morphology, but because it is modern, techy and easily funded. It is wise to recall that advances in theoretical physics have been made with no more than a piece of chalk, a blackboard and a brain, and that new does not always mean better.

In science, technology should be secondary to the pursuit of knowledge. A Lamborghini has the potential to deliver you faster than a Model-T, but only if you know your destination and are headed in the right direction. It takes courage to break from the herd and defend a 500-year-old field like morphology against popular trends. It is much easier to go along with the crowd, pretend that molecular data is superior and enjoy the spoils that come with not rocking the boat.

Having developed an inferiority complex, systematists seem vulnerable to riding the shirttails of trendy sciences. Acting as if taxonomy did not have its own mission and goals, is it any wonder that others disrespect the field and see it as a mere service? Even when funding is momentarily secured by researchers mimicking another science, it is a losing strategy. And the same is true for institutions.

DOI: 10.4324/9781003389071-17

Museums lacking the backbone to be leaders of collections-based taxonomy, following pop science bandwagons instead, deserve the second-class status they have earned. Society has vested in museums the care of collections as a research resource and permanent record of life. With this privilege comes the responsibility to grow and use collections by supporting species exploration and taxonomy.

Changing perceptions within the scientific community will not be easy. False assumptions about taxonomy are entrenched, and there are serious conflicts of interest to be overcome. Limited research dollars means that a resurgence of taxonomy and species exploration will compete with other projects. For this reason, an immediate return of support to systematics may require a public grassroots movement. It is a sad commentary that so many scientists put self-interest ahead of an obvious scientific opportunity, even when it involves a mass extinction. How the community arrived at positions antithetical to advancing knowledge is something for historians and sociologists to debate at a future date. For now, we need to respond to the biodiversity crisis with common sense, purpose and urgency.

The state of systematics is a crisis of our own making. We did not have to be unprepared to confront extinction and environmental challenges. Had we simply resumed the upward trajectory in annual species discovery after World War II, we would be in a much better position. Without knowledge of species, prospects for the best possible outcome of the biodiversity crisis are not good. Decades of marginalization, and the loss of expertise, have reduced taxonomy's capacity to describe and classify species and rendered it susceptible to Faustian bargains that promise to bring funding back to the field by compromising its principles.

You would think systematists, being historians themselves, would have learned lessons from the New Systematics era, when the mission of their science was intentionally confused with that of population genetics. Unfortunately, we are repeating the same mistake. From the 1940s until about 1970, in an effort to appear modern and share in money flowing to population genetics, taxonomists sold out the identity of their own discipline, going along with efforts to reframe systematics in the image of population genetics. Ernst Mayr's *Systematics and the Origin of Species* belittled traditional taxonomy as "bookkeeping," stressing instead the importance of "population thinking" for real scientists. Blurring the line between these disciplines undermined the identity of systematics, making it appear to be a second-rate version of population genetics.

A secret handshake for population-thinkers was the term biosystematics, chosen by Ross for his 1970 textbook. Even worse than the loss of clarity about the mission of systematics, this cultivated a mongrel science not representative of the best of either field. Population genetics is experimental, conducted in real time and focused on genetic phenomena within and between populations of living species. Taxonomy is comparative, historical, non-experimental and focused at and above the species level. The former deals with processes, the latter patterns. It is, therefore, irrational to suppose that the approach of either could prove suitable for the aims of the other. They are complementary, one telling us about micro-evolutionary processes, the other macro-evolutionary patterns. We are now repeating the same error. We are allowing systematics to be redefined, to share in the popularity of molecular genetics, at a time when its traditional aims are more important, relevant and urgent than ever. This will have a similarly disastrous outcome for taxonomy unless we reverse course, clarify our mission and see molecular techniques in their proper perspective.

Kierkegaard said that life can only be understood backwards but must be lived forwards. This is true of the course of human events and of speciation. There are many examples among isolated populations of species-in-the-making; partially diverged, but not yet distinct, populations that appear headed toward species status. Even 99% diverged, populations are not species. They are the stuff of population genetics, not systematics. To understand history, we must focus on cases where speciation—and the accompanying transformation of characters—is complete. Because species hypotheses are based on all-or-nothing claims about attributes, they can only be recognized after the fact.

Systematists bear no small measure of blame for permitting their field to be so widely misunderstood, now as then. The lack of appreciation for taxonomy has come to a head and, combined with the rapid rate of extinction, delivered us to a crossroad. Unless we have a taxonomic renaissance, its centuries-old mission, a counterbalance to general or functional biology, could be lost. We must reassert systematics as an independent discipline, clearly articulating its vision, aims, theories, methods and importance. We have traveled dangerously far down the road of replacing systematics with a pragmatic molecular approach that, in spite of its considerable strengths, is insufficient for the broader goals of systematics. The longer we delay, the more its aims are clouded and the more difficult it will be to put systematics back on course.

Misunderstood, distracted from its own goals, I fear systematics could soon enter a death spiral. The limited funding available for the study of species is disproportionately allocated to molecular projects, rote phylogenetic analyses and meeting other biologists' needs to identify species. Blindly accepted assumptions behind algorithms spit out branching diagrams that look convincing enough but make questionable assumptions about evolution and frequently ignore the full spectrum of relevant evidence. Expedient, supposedly more objective DNA analyses do not yield the same depth of understanding of species as the critical assessment of individual, complex morphological characters. With little or no knowledge of the species involved, anyone can be trained to sequence DNA and run a phylogenetic computer program. But ease comes at a high price. In the absence of revisionary taxonomy, existing hypotheses go untested, new species undescribed, names unverified and classifications unimproved.

Many molecular-based studies do not follow through to translate ideas about relationships into formal classifications. Resulting phylogenetic "trees" show putative relations among species, but none of the good stuff, the unexpected and complex evolutionary novelties that make species interesting. It is reasonable to question the value of being able to identify species about which we know almost nothing. And it is appropriate to question why phenetics, that is, measures of overall similarity, should be accepted in molecular studies when it was firmly discredited by morphologists decades ago.

An alarming trend away from deep scholarship and toward the expediency of molecular surrogates is clear, as noted by John James Wilson: "Species identifications have become 'DNA barcoding,' new species discoveries are characterized by genetic divergences, and traditional classification has been supplanted by molecular phylogenetics." During what may be our last opportunity to understand species diversity, failure to describe them, to document what makes each unique and fascinating, is inexcusable. Automated DNA sequencers and off-the-shelf software seem to put species identification and phylogenetic analysis within the reach of anyone, but at the expense of expertise, theoretical depth and the most valuable knowledge. A calculator in the hands of a child may solve an equation, but does not a mathematician make.

Taxonomy-as-cookbook, slavishly following protocols, is a hollow shell of the intellectually rich and challenging traditions of systematics. Insights gained through comparative morphology and scientific natural history observations can be augmented, but never matched, by molecular data. Much of the fight has been beaten out of the dwindling number of professional taxonomists, after decades of being denied recognition and funding. Making matters worse, those promoting a molecular-based taxonomy falsely frame the situation: "… the old museum specimens have little value, because of the enormous cost in time and budgets to examine and identify them morphologically" (Sharkey et al.). This ignores the vastly greater amount of information associated with specimens in collections compared to DNA sequences. It uses the lack of past investments to make morphology-based studies faster and more efficient as an excuse to continue neglecting morphology. Applied to literature, their argument is that we should be content to look at pictures in Dr. Seuss' books because it takes too long to read *War and Peace*. I guess it depends on how much you wish to learn. My hope is that a clear-throated, forceful reassertion of what makes systematics fascinating, combined with public recognition of its importance, will give renewed heart to taxonomists. And that we begin to inspire and educate a new generation of specialists devoted to a deep knowledge of taxa rather than superficial molecular studies.

I recently watched a fascinating public television program on the evolutionary history of crocodiles that, among other things, gave DNA a leading role in the narrative. Viewers were left with the impression that we had lived in backward ignorance about evolutionary relationships until DNA arrived to save us. Without detracting from impressive contributions by DNA, the elevation of molecular data to savior status is a bit much. In this case, DNA is only able to shine because of past contributions from taxonomy, morphology and paleontology. Showing up at the 11th hour to rescue wretched taxonomists from themselves is possible only because so much reliable ground work had been laid by traditional means. Recognizing that no such descriptive background exists for 80% of species, DNA will have precious little to offer we retrogrades unless someone describes morphology, too.

Evidence of the assault on systematics is in plain sight. Natural history collections are physically cared for by staff, but their growth and development are not the same top priorities they once were for leaders of museums. The relationship between revisionary taxonomy and collections is symbiotic. In the absence of active taxonomic research, the amount and reliability of information attached to specimens degrade; and, when the expansion, development and curation of collections are neglected, it becomes an obstacle to systematics research. Researchers hired by museums increasingly engage in a variety of popular science projects that have little or nothing to do with collections. Although anecdotal, I am aware of an increasing number of prestigious endowed chairs in American universities, established to support taxonomy, now occupied by faculty with no interest or competence in systematics. When systematists are not hired, museums fail to use collections for their most impactful purposes and universities are not preparing the next generation of taxonomists.

Many graduate programs that used to emphasize monographs, revisions and floras have been eliminated or diluted beyond recognition. Textbooks and curricula give less and less attention to either taxonomy or natural history generally as noted by Dijkstra:

> Within a 40-year period in the United States, textbook content related to natural history decreased from two pages of every three to just one page; related PhDs fell from two in five to one in five (even as the total number in biology tripled); and the median number of courses on natural history required for a biology bachelor's degree dropped from two to zero.

Grants for descriptive studies are virtually nonexistent, unless they are sold as a molecular project. No such expectation exists for DNA projects that are free to ignore morphology entirely or relegate it to a literature review. In the U.S. and Europe, few systematists are hired with the expectation of conducting morphology-based taxonomy. At best, it must be disguised as some more respectable research or pursued as a sideline. Given the rate of extinction and how little we know of life on earth, this is insanity.

There was a time when every self-respecting university biology program had taxonomists on faculty. Taxonomists bring a uniquely broad perspective to the classroom, rich in details about species, evolutionary history, biogeography and natural history. But the commitment to the breadth of expertise has fallen away as universities prioritize external funding above academic balance. Faculty impact is measured in research dollars and citation indices rather than advances in knowledge. A paper describing the latest molecular technique may receive thousands of citations, while a taxonomic monograph of exquisite scholarship, which will remain an important reference for generations, receives few immediate citations and is therefore considered of little value. Excellence takes a back seat to popularity, scholarship to money and systematics to fads.

Most scientific publications are overtaken by subsequent studies within a few years and rarely, if ever, cited thereafter. Taxonomic treatments suffer in this beauty pageant for two reasons. First, the narrow focus on one group, such as a genus, means that there are few readers at the time of publication, even though many will consult the document over its long shelf life. Second, taxonomic literature is shamefully and unethically ignored in citations. Ecologists routinely rely on taxonomic literature to identify species without attribution. In contrast, every species named since 1758 (1753 for plants, 1757 for spiders) must be accounted for in subsequent taxonomic papers. Primary taxonomy papers are rarely cited outside of the systematics community. Instead, information is translated into user-friendly secondary books, field guides and web sites that are accessed routinely by field biologists who assign absolutely no credit to taxonomists who created the knowledge. Ecologists work with significant numbers of species, so the primary sources ignored are significant. In a just world, taxonomic papers would be cited each time that species names are used. Medical publications, for example, would cite Linnaeus, 1758, who named both their primary and literal "guinea pigs," *Homo sapiens* and *Cavia porcellus*. It is academic malpractice to use taxonomists' intellectual property without acknowledgment. But such plagiarism is universally tolerated contributing to the perception that taxonomy really isn't all that important.

A flora or fauna is a study of species in a specified geographic area, such as the plants of Panama or insects of the Great Smoky Mountains National Park. A revision is a study that reviews every species belonging to a higher taxon, such as a genus or family, regardless of where on earth they are found. The ultimate taxonomic publication is a monograph, a revision that goes the extra mile to summarize much or all of what is known of the taxon, in addition to describing and validating species, making them identifiable and classifying them to reflect relationships. A fine example is Herman Lent and Pedro Wygodzinski's monograph of triatomine bugs, vectors of Chagas disease in Central America, which includes a thorough overview of these medically important insects.

Revisions and monographs remain the gold standard in systematics. Their comprehensively comparative approach makes them unmatched in efficiency, too. The reason is simple. In order to test the validity of a single species, it is not only necessary to compare hundreds or thousands of specimens from throughout its geographic range, but to examine large numbers of specimens of related species, too. This is an enormous amount of work to assess the boundaries and status of a single species. However, when specimens of every species in a group are compared at the same time, it becomes quite efficient by avoiding redundancy. What is learned about each species is immediately applicable to others.

When we plot the taxonomic learning curve, the discovery and naming of species through time, we typically see a graph with a slowly rising slope interrupted by occasional spikes. These spikes correspond to revisions and monographs. A few species are named between, but it is during a revision, when backlogged specimens in museums are sorted, known species tested (and sometimes split, or sunk) and new ones named. When Kelly Miller and I revised the beetle genus *Agathidium*, it was only the third comprehensive review of North and Central American species. The first was by George Horn in 1880. Dr. Horn was a Philadelphia obstetrician who once commented that the stork tended to come at night, happily leaving his days open for studying beetles. I wonder how his patients would have reacted, knowing where his hands had been. The second was by Henry Clinton Fall in

1934, who approximately doubled the number of species known to Horn. Our revision was published in 2005, naming 65 new species, once again doubling the number known. We are aware of several areas not yet adequately collected for the genus, including the southern Appalachians, the Ozarks, the highlands of Mexico and Central America and perhaps the Pacific Northwest. There remains much for the next reviser to accomplish, and it is conceivable that the number of these beetles could be doubled yet again before we approach full knowledge of New World *Agathidium* species.

With so many species and so few taxonomists, revisions and monographs have always been published infrequently, often just once or twice per century. The problem this creates, of course, is that revisions are out of date as soon as the next new species is found. There are obvious steps that could be taken to maintain up to date information. It is easy to see the benefits of electronic revisions, or e-monographs, which are dynamic, incorporating the latest information and providing near-real-time access to all that is known of a group. It is, therefore, unfair for those who deny funding to modernize taxonomy to also criticize out-of-date knowledge.

Dozens of initiatives have claimed to address the so-called taxonomic impediment, our inability to identify species, with little attention to the needs of taxonomists. Such token efforts are not enough. Improving the ease with which users can access information does nothing to assure that information is complete, verified and current. The simple fact is that the situation will not improve until we return support to systematics.

What is needed is a new generation of tools designed by and for taxonomists to do taxonomy faster and better. Reliable information on web pages should be a by-product of vibrant taxonomic research. To ignore the needs of taxonomists does a great disservice to science. It stifles the advance of our understanding of species and evolutionary history and denies users access to the highest quality information. Simply relocating outdated and incomplete information to the Internet is not nearly enough. We must not measure progress by how easy it is for users to access taxonomic information, or by the number of species recognized by DNA barcodes, but by the rate at which descriptive knowledge of species is created, tested and disseminated. Because other biologists simply want to identify species, they may not appreciate why descriptions are so important. To understand species in an evolutionary context, however, is to understand their characters, too.

Systematists deal with lots of predictions. Every individual in a species is predicted to possess a specified combination of characters not present outside the species. Relationships among species are indicated by synapomorphies, another prediction. And synapomorphies are based on hypotheses about homology. A monophyletic group is hypothesized to include an ancestral species and all, and only, its descendants. Stated in terms of Occam's razor, it is simpler to initially assume that an attribute shared by species was inherited from a common ancestor rather than evolved separately. There are exceptions, of course. Convergent evolution does take place. This makes it important to examine as many relevant characters as possible so that the distribution of a convergence or reversal is seen as conflicting with a pattern common to other data.

The prediction that all vertebrate animals share the character "backbone" has yet to be falsified, holding true after examining millions of specimens representing 70,000 vertebrate species. But sharing the character backbone does not imply that backbones all look alike. The backbones of guppies, geckoes and giraffes are different at a glance. Close inspection, however, reveals that they are built of the same parts and, in spite of differences in size, shape and detailed conformation, each is an example of one and the same character.

The theoretical framework of homology—the idea that differing attributes can nonetheless be interpreted as variations on one and the same body part—was advanced by the British comparative anatomist Richard Owen, one of my personal heroes. The term would later be explicitly redefined in the context of evolution, ascribing homologs to common ancestry. I used to walk past a statue of Owen on the way to my office in the Natural History Museum in London, reminding me of disagreements among historians as to Owen's opinion on evolution. Publicly, at least, he was a vocal opponent of Darwin's views. Yet, his concept of homology makes it difficult to dismiss the possibility that he was a closet evolutionist. A product of his time, it is conceivable that he was indeed a creationist. On the other hand, he may have simply been a remarkably astute politician. Members of Parliament in his day were educated in Oxford, or another university, with a curriculum designed to prepare the clergy. He dazzled people with his ability to infer the habits of extinct species from the structure of a few bones, assumed by many as evidence of Devine design. This seeming affirmation of a Creator played no small part in Owen's success in securing government funding to engage Waterhouse to design the Natural History Museum in South Kensington, certainly a strong potential motive to oppose Darwin publicly.

Step by step, everything that makes systematics unique and intellectually rewarding is being eliminated or watered down. Thinking deeply about individual characters is becoming a lost art, replaced by automation, simply dumping masses of raw data into a computer program to be analyzed as if all data were equal. This move from critical thinking to technology is a risk in itself, increasingly supplanting ideas and theories with procedures and protocols. I'm not opposed to technology, only technology masquerading as science.

It is imperative that we allow taxonomists to return their focus to species, characters and relationships, and to produce descriptions, classifications and revisions. Deep knowledge of characters leads to unexpected evolutionary and natural history insights. Asking only to tell species apart, it is possible to miss the most fascinating things about them. Molecular and morphological data are a powerful duo but divorced from morphology, DNA dumbs down taxonomy. And unlike theories behind the interpretation of complex morphology, a DNA sequence is merely a description of the order of nucleobases. The very kind of "mere" description that morphology is falsely accused of being.

Instead of simply incorporating molecular data into systematics, it has grown out of proportion. Perceived as new and technologically sophisticated, basking in the glow of its popularity in genetics and biomedicine, the increasingly dominant role of molecular data in systematics is something one dare not question in civilized company What should have been an exciting addition to taxonomy is instead taking center stage and putting the mission of systematics at risk. If this trend continues, the goals of systematics may soon be replaced or changed beyond recognition.

Years ago, when DNA sequencing was new, I was invited to interview for a vice presidential position at a leading natural history museum. I was asked about the merits of establishing a department of molecular systematics. I said it was a terrible idea, adding that DNA is data, not a discipline. It would be like, I opined, creating a department of compound microscopy, and that would not be very intellectually interesting. My position on the important but limited role of DNA in systematics has only hardened since. And I reject the uncritical elevation of molecular data above other evidence. Comparing the information content of DNA to that of morphology, I'm convinced its popularity has more to do with funding and the adulation of other biologists than a genuine conviction about the superiority of DNA data.

From the perspective of systematics, molecular data seems most important at two extremes. At a most basic level, molecular data is key to understanding homologous characters, transformations of embryonic developmental pathways and bridging genomes and the expression of complex morphology. At the applied extreme, it can provide identification tools that can be used by field biologists to disambiguate species. Between the two, however, are theories, hypotheses and descriptions having to do with characters, species and clades that are most interesting and rewarding when morphology and other relevant evidence are synthesized.

Through much of my career, I have watched as the centuries-old discipline that explores and classifies species was misunderstood, maligned and marginalized. Now, the popularity of molecular data threatens to further diminish this theory-rich, intellectually exciting science by shifting its focus to meeting short-sighted, immediate needs. This would be a great injustice any time, but to neglect systematics and its unique mission in the middle of a biodiversity crisis is inexcusably irresponsible. There is far too much at stake, too many things we do not know, things that can only be learned by practicing real systematics. At the same time, society is being overtaken by environmental challenges unlike any in human history: accelerated species extinctions, degradation of ecosystems, conversion of wilderness, pollution, a deluge of solid waste and a reliance on overexploited and nonrenewable resources for our prosperity among them. Effective responses depend on basic knowledge of species that we do not have, and which DNA cannot deliver by itself.

The reasons are as simple as they are self-evident. If we do not know what species exist, or where they live, how can we detect changes in biodiversity, monitor the rate of extinction, guard against invading pests or develop measurable conservation goals? If we do not know what makes each species unique and interesting, how can we value it as an individual kind? If we allow large numbers of species to go extinct before they are discovered, how can we hope to ever understand the origins and history of biodiversity and all that makes us human? And, if we throw away the opportunity to explore how other species have adapted to live efficiently, where else can we turn for clues and inspiration to create a sustainable future for ourselves?

The traditional mission and goals of systematics are being replaced by those of molecular genetics. I have no objection to supporting molecular geneticists to pursue their own goals, but I resent taxonomy being hijacked. It is indisputable that DNA is a valuable tool for taxonomists, a tool that should be integrated with traditional sources of evidence. But DNA sequences are being treated as if they were infallible revelations of truth. In fact, they are no better, and for many purposes are worse, than other sources of evidence. The mad rush to replace traditional data with molecular is creating more problems than it is solving, and wasting valuable time. As others have observed, what is good about DNA data is not new, and what is new is not good.

Ole Seberg has sounded the alarm with respect to our understanding of evolution:

> To relegate taxonomy to a high-tech service industry centered around a few DNA sequences will deprive evolutionary biology of its most important function: the testing of evolutionary hypotheses at all levels from the evolution of characters, over the evolution of species, to the evolution of clades, i.e., species concepts, species delimitations, phylogenetic reconstructions, homology statements, character polarizations, and ultimately classifications are all scientific hypotheses that do not hinge upon a few DNA data points, but change as science progresses.

I do not want to leave the impression that no one cares, or that tentative steps are not being taken by concerned individuals, organizations and institutions to address the plight of systematics. No doubt well-intentioned, most of the efforts to reinvigorate taxonomy in recent years have failed for two reasons: they have misunderstood the mission and goals of taxonomy and they have focused on the needs of consumers of taxonomic information, ignoring the requirements of taxonomists to do taxonomy and increase, improve and verify the very information they seek. The NSF is to be commended for its long-term support of systematics. Perhaps, 95% of competitive funding for the field in the U.S. comes from this one government agency. But to exact the greatest impact with its limited budget, the NSF has understandably chosen to prioritize projects that advance the field as a whole over those that simply move a limited body of empirical knowledge forward. Projects that test theories, perfect techniques and develop general research resources outcompete a straight forward revision of a genus, for example. It is rational for the NSF to question why it should fund the study of any one taxon at the exclusion of other, equally deserving ones. The problem is that those engaged in the heavy lifting of testing 2 million existing species hypotheses, and the initial discovery and description of millions of new species, have nowhere else to turn for funds to support what should be routine inventories and monographs. Because it is nearly impossible to fund such basic, descriptive work, institutions shy away from hiring scientists interested in such studies. For reasons of finance and profile, they prefer to hire faculty capable of bringing in large grants in more fashionable areas of science. It is beyond question that such monetary interests are a major driver behind the runaway popularity of DNA studies.

Fortunately, because taxon experts care so deeply about the organisms they study, they refuse to give up. In part, because many traditional scientific journals do not want to publish taxonomy, innovative online journals, like *Zootaxa* and *Phytotaxa*, are publishing record-setting numbers of papers, and taxonomists continue to name nearly 20,000 new species each year. But this rag-tag army of determined scientists is not nearly enough to get the job done. We need a clear vision and plan, a reliable source of funding, international organization and a few bold leaders among natural history museums, botanical gardens and research universities to hire taxonomists to do taxonomy and make a long-term commitment to the completion of a species inventory. The scale of doing an inventory of life on an entire planet is so large that government support is necessary in the form of a multinational consortium. All the same, philanthropists who want to break from the pack, have an enormous impact, support science that is truly groundbreaking and enable contributions of unparalleled importance to our welfare and intellectual lives, could do no better than to get behind systematics, a fundamental field that has been orphaned by most sources of science funding.

An equally disturbing lack of perspective is evident in the ecology community. It too has been adulterated by money. Increasingly, ecology sees little value in biodiversity beyond what it can do for people, such as ecological services and natural resources. This is a sad, cynical view of the world, a view that could ultimately lead us into a future of nothing but artifice: human-designed landscapes and monocultures where wilderness once flourished. As president of an environmental college that included a leading department of landscape architecture, I came to deeply respect the mission and accomplishments of that discipline. We will be well served to seek their vision and leadership as we conceive better human-built spaces, but it would be a calamitous mistake to cede to them, or anyone other than Mother Nature, the design of the entire biosphere. The best plan involves saving enough wilderness so that it can care for itself while closing the distance between humans and the natural world.

Further Reading

Abrahamse, T., Andrade-Correa, M. G., Arida, C., et al. (2021) *The Global Taxonomy Initiative in Support of the Post-2020 Global Biodiversity Framework*, CBD Technical Series No. 96, Secretariat of the Convention on Biological Diversity, Montreal, 103 pp.

Adamowicz, S. J., Boatwright, J. S., Chain, F., et al. (2019) Trends in DNA barcoding and metabarcoding. *Genome*. doi:10.1139/gen-2019-0054

Agnarsson, I. and Kuntner, M. (2007) Taxonomy in a changing world: Seeking solutions for a science in crisis. *Systematic Biology*, 56: 531–539.

Bebber, D. P., Wood, J. R. I., Barker, C., and Scotland, R. W. (2013) Author inflation masks global capacity for species discovery in flowering plants. *New Phytologist*, 201: 700–706.

Buyck, B. (1999) Taxonomists are an endangered species in Europe. *Nature*, 401: 321.

Carpenter, J. M. (2003) Critique of pure folly. *The Botanical Review*, 69: 79–92.

Cognato, A. I., Sari, G., Smith, S. M., et al. (2020) The essential role of taxonomic expertise in the creation of DNA databases for the identification and delimitation of southeast Asian ambrosia beetle species (Curculionidae: Scolytinae: Xyleborini). *Frontiers in Ecology and Evolution*, 8. doi:10.3389/fevo.2020.00027

Coleman, C. O. (2015) Taxonomy in times of the taxonomic impediment—Examples from the community of experts on amphipod crustaceans. *Journal of Crustacean Biology*, 35: 729–740.

Costello, M. J., May, R. M., and Stork, N. E. (2013) Can we name Earth's species before they go extinct? *Science*, 339: 413–416.

Crisci, J. V. (2006) One-dimensional systematists: Perils in a time of steady progress. *Systematic Botany*, 31: 215–219.

Dijkstra, K.-D. B. (2016) Natural history: Restore our sense of species. *Nature*, 533: 172–174.

Godfray, H. C. J. (2002) Challenges for taxonomy. *Nature*, 417: 17–19.

Hopkins, G. and Freckleton, R. P. (2002) Declines in the number of amateur and professional taxonomists: Implications for conservation. *Animal Conservation*, 5: 245–249.

House of Lords (2002) What on Earth? The threat to the science underpinning conservation. Select Committee on Science and Technology, 3rd report (HL Paper 118), London.

Jones, B. (2017) A few bad scientists are threatening to topple taxonomy. *Smithsonian Magazine*, 7 September.

Kaplan, S. (2022) As many as one in six U.S. tree species is threatened with extinction. *The Washington Post*, 23 August.

Kierkegaard, S. (1843) *Søren Kierkegaards Skrifter*, Søren Kierkegaard Research Center, University of Copenhagen, Copenhagen 1997, vol. 18, p. 306.

Kirby, W. and Spence, W. (1828) *An Introduction to Entomology*, vol. 1, Longman, Rees, Orme, Brown, and Green, London, 517 pp.

Krell, F. (2000) Impact factors aren't relevant to taxonomy. *Nature*, 405: 507–508.

Lagomarsino, L. P. and Frost, L. A. (2020) The central role of taxonomy in the study of Neotropical biodiversity. *Annals of the Missouri Botanical Garden*, 105: 405–421.

Lambertz, M. (2017) Taxonomy: Retain scientific autonomy. *Nature*, 546: 600.

Lammers, T. G. (1999) Plant systematics today: All our eggs in one basket? *Systematic Botany*, 24: 494–496.

Landrum, L. R. (2001) What has happened to descriptive systematics? What would make it thrive? *Systematic Botany*, 26: 438–442.

Lipscomb, D., Platncik, N., and Wheeler, Q. (2003) The intellectual content of taxonomy: A comment on DNA-taxonomy. *Trends in Ecology and Evolution*, 18: 65–66.

Lee, M. S. Y. (2000) A worrying systematic decline. *Trends in Ecology and Evolution*, 15: 346.

Lent, H. and Wygodzinsky, P. W. (1979) Revision of the Triatominae (Hemiptera, Reduviidae), and their significance as vectors of Chagas' disease. *Bulletin of the American Museum of Natural History*, 163: 125–520.

Mace, G. M. (2004) The role of taxonomy in species conservation. *Philosophical Transactions of the Royal Society of London, B*, 359: 711–719.

Marder, J. (2014) As species decline, so do the scientists who name them. *PBS News Hour*, 7 May.

May, R. (2004) Tomorrow's taxonomy: Collecting new species in the field will remain the rate-limiting step. *Philosophical Transactions of the Royal Society of London, B*, 359: 733–734.

Mertl, M. (2002) Taxonomy in danger of extinction. *Science*, 22 May. www.science.org/content/article/taxonomy-danger-extinction

Miller, K. B. and Wheeler, Q. D. (2005) Slime-mold beetles of the genus *Agathidium* Panzer in North and Central America. Part II. Coleoptera: Leiodidae. *Bulletin of the American Museum of Natural History*, 291: 1–167.

Newton, A. (2021) One-third of the world's tree species are threatened with extinction—Here are five of them. *The Conversation*, 15 September.

Pearson, D. L., Hamilton, A. L., and Erwin T. L. (2011) Recovery plan for the endangered taxonomy profession. *BioScience*, 61: 58–63.

Pinheiro, H. T., Moreau, C. S., Daly, M., and Rocha, L. A. (2019) Will DNA barcoding meet taxonomic needs? *Science*, 365: 873–874.

Rawlence, N. (2018) Taxonomy, the science of naming things, is under threat. *The Conversation*, 13 November.

Sangster, G. and Luksenburg, J. A. (2015) Declining rates of species described per taxonomist: Slowdown of progress or a side-effect of improved quality in taxonomy? *Systematic Biology*, 64: 144–151.

Schuh, R. T. (2003) The Linnaean system and its 250-year persistence. *The Botanical Review*, 69: 59–78.

Seberg, O. (2004) The future of systematics: Assembling the tree of life. *The Systematist*, 23: 2–8.

Seberg, O., Humphries, C. J., Knapp, S., et al. (2003) Shortcuts in systematics? A comment on DNA taxonomy. *Trends in Ecology and Evolution*, 18: 63–65.

Sharkey, M. J. Janzen, D. H., Hallwachs, W., et al. (2021) Minimalist revision and description of 403 new species in 11 subfamilies of Costa Rican braconid parasitoid wasps, including host records for 219 species. *ZooKeys*, 1013: 1–665.

Tewksbury, J. J., Anderson, J. G. T., Bakker, J. D., et al. (2014) Natural history's place in science and society. *BioScience*, 64: 300–310.

Thiele, K. and Yeates, D. (2002) Tension arises from duality at the heart of taxonomy. *Nature*, 419: 337.

Wagele, H., Klussmann-Kolb, A., Kuhlmann, M., et al. (2011) The taxonomist—an endangered race. A practical proposal for its survival. *Frontiers in Zoology*, 8: 25.

Wheeler, Q. (2004) Taxonomic triage and the poverty of phylogeny. *Philosophical Transactions of the Royal Society of London, B*, 359: 571–583.

Wheeler, Q. (2008) Introductory: Toward the new taxonomy. In *The New Taxonomy* (ed. Q. Wheeler), CRC Press, Boca Raton, 237 pp.

Wheeler, Q. (2013) Are reports of the death of taxonomy an exaggeration? *New Phytologist*, 201: 370–371.

Wheeler, Q. and Cracraft, J. (1997) Taxonomic preparedness: Are we ready to meet the biodiversity challenge? In *Biodiversity II* (eds. M. L. Reaka-Kudla, D. E. Wilson, and E. O. Wilson), Joseph Henry Press, Washington, 435–446.

Wheeler, Q. D. and Miller, K. B. (2005) Slime-mold beetles of the genus *Agathidium* Panzer in North and Central America. Part II. Coleoptera: Leiodidae. *Bulletin of the American Museum of Natural History*, 291: 1–167.

Wheeler, Q. D., Raven, P. H., and Wilson, E. O. (2004) Taxonomy: Impediment or expedient? *Science*, 303: 285.

Wilson, J. J. (2011) Taxonomy and DNA sequence databases: A perfect match? *Terrestrial Arthropod Reviews*, 4: 221–236.

16 The Nature Gap

In a word, our relationship with nature is broken. From urbanization to social media, a worrisome, widening gap exists between humans and the natural world. Each of us would do well to spend more time in the company of other species. Whether keeping fishes in aquaria, tending plants in a garden, hunting or hiking in wilderness, connecting with other species enriches our lives. It reminds us that we are neither alone in the world, nor completely in charge. Beyond our built environments are wills other than our own and forces beyond our control.

To make sure that our conservation ethics are more than theoretical, it is important that we meet some of the vulnerable species whose fate we hold in our hands. This was the magic in Elizabeth Kolbert's *The Sixth Extinction*, seeing endangered species as individual, precious beings. Cold statistics tell one story but are a poor motivator for making sacrifices required for conservation. And selfish attention to our own needs, focusing on the products and services derived from ecosystems, distracts us from paying attention to the plants and animals involved. Being up close and personal with species is important, but not enough. Neither is understanding their roles in the web of life. To fully appreciate our ties to other species requires us to see them, and ourselves, in the context of phylogenetic history. Some people are afraid of what they might see if they look too closely into the evolutionary mirror. In so doing, they can no longer pretend that humans are separate from the beasts, much less superior to, or even more highly evolved, than other species. On the contrary, we are merely different … just like every other species. History binds *Homo sapiens* to every other species that is or has been part of the amazing story of evolution. In spite of our haughtiness, humans are one terminal twig among millions on the tree of life, each of our human attributes traceable to precursors found in ancestors.

A scientific inventory of species is a necessary step toward a framework of understanding but cannot substitute for personal experience. Rather than an end, an accounting of species is a beginning, an invitation to lace up your boots, hit the trail and experience biodiversity for yourself. To not only take time to smell the roses, but also to observe their pollinators as they come and go; to not simply enjoy a day at the beach, but spend a few hours getting to know the invertebrates stranded in a tidal pool.

DOI: 10.4324/9781003389071-18

Deeply wired in our psyche are connections to other species that E. O. Wilson described as biophilia. It's good that this innate tie to other living things resides in our subconscious, even better when we bring it to the fore. We owe it to ourselves to get back in touch with the natural world, to make the acquaintance of species we've never met and to immerse ourselves in places so wild and vast that we are made to feel small. In wilderness, no longer in control of our surroundings, we are reminded that the biosphere is made up of an enormous number of species and that we, in spite of our cleverness, awesome power and sense of self-importance, are only one among many with equal claim to call this planet home.

As human populations become increasingly urban, experiences with wild species and places are becoming less common for millions of us. As we spend more hours surfing the Web or engaged with social media, or, come to think of it, writing a book, we have less time for direct interactions with other species or, raising a different set of problems, other people. Sensing that our needs are being met by human-constructed systems, the intuitive understanding that our lives are intertwined with, and dependent upon, those of other species is fading. Diverse species are becoming things we watch on nature programs or read about in magazines rather than beings we experience for ourselves. Divorced from intimate contact, it is much easier to accept species extinctions as unfortunate events that have little direct impact on our daily lives.

Connections with species are important to the human spirit. Stepping outside a tightly controlled built world, we touch something primordial in our constitution. We know that close interactions with animals can be comforting, even therapeutic, as evidenced by service canines. Perhaps it's their nonjudgmental, unconditional love. There is no better place for deep contemplation than the solitude found in wilderness; nothing better to restore humility than being reminded that natural systems exist with great powers over which we have no control; and nothing to put us in our place faster than recalling that the biosphere existed for billions of years before humans arrived and will persist, in one form or another, long after we've gone.

Unless we reconnect with other species, the future for many of them is not promising. President Theodore Roosevelt understood this in his gut, a result of his immersive relationship with wilderness. Many find it abhorrent that Roosevelt would shoot hundreds of grouse in one day, finding his hunting habits antithetical to conservation. But the truth is exactly the opposite. He loved and worked to protect other species, including the creation of the national park system, precisely because of his personal, if sometimes predatory, association with them. It matters less whether our view of other species is through the lens of a camera or sight on a rifle, so long as we are looking.

I once did field work in a national forest where children arriving on school buses were instructed that they could remove nothing, not even a leaf. No doubt well-intentioned, this extreme *laissez-faire* attitude is destined to backfire. Preserving nature under a bell jar, turning it into something you look at, but cannot touch, will not kindle the kind of connection between people and nature necessary if they are to make sacrifices to save other species. Conservation goals are far better served by buying children plant presses and butterfly nets and encouraging them to go forth, collect and discover.

Making a personal herbarium or insect collection is an ideal hobby for child or adult. You have the satisfaction of mastering a complex body of knowledge, a sense of exploration as you search out species you have not previously seen and assembling a collection is paid for more in labor and studious hours than dollars.

As the human population grows, and many species and habitats are imperiled, collecting specimens in nature becomes controversial. It is unethical to collect a rare orchid for personal pleasure or dump barrels of rotenone into a mangrove swamp to capture a few fish for the aquarium trade. But the ethics of collecting are not so simple. There are perfectly valid reasons for professional botanists to collect rare orchids. Without the knowledge they create, we would be in a far worse position to evaluate the status of species and inform conservation efforts. And rotenone can be a useful tool in the hands of ichthyologists for both assessing the diversity and status of fishes.

If a large number of amateur collectors descended on the habitat of a rare species, it could be driven to extinction in short order. Some species have small, vulnerable populations and their collection should be strictly limited. There are scientific justifications for collecting threatened species, but they must be damned good ones. Highly restrictive rules that make sense for rare plants and vertebrate animals, however, make no sense for other groups, such as protists in a vernal pond or midges at a porchlight.

There are some species of insects with small populations that are no less vulnerable than any vertebrate, but this is not so for others. Many insects are widespread and abundant, and there is no reason to think limited collecting presents any threat to them. I am amused when people denounce entomologists for collecting insects, then call an exterminator at the first sight of a bug in their home. You can't have it both ways. All it takes is one infestation of head lice to convert the staunchest opponent of killing insects. I love and respect insects and have never taken pleasure in killing them. That said, their small size and multitude of species mean that one cannot study them adequately without collecting specimens and getting them under a microscope. And unless specimens are preserved in natural history collections, there is no way to repeat or verify earlier observations. The fact is, an astronomical number of insects are killed each summer by automobiles and, unless you are prepared to give up modern travel, you have no moral high ground on which to oppose judicious scientific collecting.

Nature should not be something you only read about in books; it is best experienced firsthand. There is nothing like a few days in a lowland tropical rainforest to appreciate other species, but it is not necessary to go so far afield. As Yogi Berra said, you can observe a lot by just watching. Few people have spent quality time closely examining a decaying log. Peel back the bark, roll the log over or split it open and all kinds of multilegged biodiversity go scattering. One can see fungi and slime-molds they never noticed before, and a diversity of arthropods, from woodlice to wood-boring beetle larvae. The key is personally experiencing species you have never seen up close. This direct interaction with species motivates birders to travel, but I guarantee there are species much closer to home that are worthy of your attention.

When I was about eight years old, a friend received a department-store microscope for his birthday. A toy with chrome knobs and cheap lenses changed my life. We had examined all we could get our hands on: the prepared slides that came with it, petals off a dandelion, a bit of skin peeled from my palm. Then it happened. We took a drop of stale water from a puddle and placed it on a slide. Frenzied shadowy figures were darting back and forth. As I slowly turned the knob, a world of microscopic species came into focus that I had no idea existed. As ciliate protists and rotifers raced in and out of view, I could not have been more astonished were I an astronaut seeing extraterrestrials on some other planet.

I spent much of my childhood searching out and culturing as many protist species as I could. By middle school, I was producing pure cultures of species like *Paramecium caudatum* by the gallon. Countless hours were spent observing protists swim, engulf and divide. But the greatest thrill was seeing species I had not seen before. Buying a copy of Kudo's college textbook, I was able to identify a wide range of ciliates, flagellates and sarcodines. My discoveries made such an impression on me that, to this day, I could take you to the exact ponds and ditches where I first collected *Volvox globator*, *Euglena gracilis* and *Lacrymaria olor*.

My interest shifted to insects when I got to college, but my passion for species exploration never waned. I remain in awe of species diversity, spellbound by the story of their evolution as it unfolds before me through comparative morphology studies. Not everyone will become a taxonomist like I did. But such intimate interactions with species instill a deep sense of wonder, a respect for nature that cannot be had any other way. Anyone experiencing a fraction of the excitement I have enjoyed is destined to have a deep regard for other species and to be a supporter of conservation efforts, natural history museums and actions that protect the natural world.

In *Biophilia*, Ed Wilson addressed the apparent tension between conserving species and leaning on nature to meet the seemingly boundless needs and wants of humanity:

> Natural philosophy has brought into clear relief the following paradox of human existence. The drive toward perpetual expansion—or personal freedom—is basic to the human spirit. But to sustain it we need the most delicate, knowing stewardship of the living world that can be devised. Expansion and stewardship may appear at first to be conflicting goals, but they are not. The depth of the conservation ethic will be measured by the extent to which each of the two approaches to nature is used to reshape and reinforce the other. The paradox can be resolved by changing its premises into forms more suited to ultimate survival, by which I mean protection of the human spirit.

As we engage in biomimicry with greater purpose and success, relying on diverse species for inspiration, Wilson's point will be reinforced. We will see that in protecting and learning from species we protect also the human spirit, giving ourselves the wherewithal to continue to indefinitely innovate and expand our horizons.

At the human level, a movement in the interior design world reinforces our subconscious connection to other species. William Browning and Catherine Ryan explain that nature-based motifs integrated into places where we live and work are comforting, lowering blood pressure, reducing stress and contributing to a general sense of well-being. Hospital patients recover faster with a view of trees rather than a brick wall. Students perform better academically with even a hint of nature. It is ironic that we are discovering benefits from the mere visual suggestion of other species at the same time that we drive thousands of plants and animals to extinction. With the extinction of species goes the possibility of experiencing them for real, a point not lost on Browning and Ryan: "While bringing nature inside can help to create meaningful and enduring experiences of the built environment, ultimately the most important suggestion we can make is to get outdoors and directly engage with nature." Amen to that.

John Muir said that in every walk with nature, one receives more than he seeks. That has been my experience, too. For those willing to see it, overwhelming beauty is found in nature at every scale: from minute morphological structures, such as diffraction gratings making an insect iridescent, to entire landscapes, like a mature forest or coral reef. It is impossible to deny the good and healing that intimate contact with nature brings. Cornell botanist and educator Liberty Hyde Bailey astutely observed that there is nothing wrong with a wayward juvenile that cannot be fixed between the handles of a plow. Close interaction with other species, even those tamed in agricultural settings, puts us back in touch with our roots. Seeing the cycle of life on a farm, or in the wild, reminds us of our mortality and our ephemeral place in the long trajectory of life.

Most taxonomists I know find field work restorative. It's a chance to leave the worries and diversions of everyday life behind, unclutter your mind, soothe your soul and recharge your mental batteries. The adventure of field work, the exhilaration of discovery and the peacefulness of communing with wilderness have no equal. Something is lost when this rich intellectual, emotional and sometimes spiritual experience is traded for the cold efficiency of a sterile laboratory. In a very real sense, something is lost if species are kept at a technological arm's length. Taxonomy includes many impersonal activities, of course, such as literature searches and plotting dots on distribution maps. These have a place in science, but do not substitute for focusing on what really matters: species themselves, their amazing attributes and their places in the biosphere and history.

We would do well to expose every child to diverse species, letting them discover which kinds of organisms speak to them. Like falling in love, there is no way to know in advance which combination of attributes of species, or a potential life partner, we will find irresistible. Some will be drawn to particular species, like dogs or horses. Others to the amazing diversity within some higher taxon, perhaps a family of flowering plants or moths. In every case, their appreciation for nature, empathy for other species and understanding of themselves will grow.

Exploring species awakens a curiosity and sensitivity for nature in children and elicits a state of childlike wonder in adults. So, get out and enjoy species. Go for a hike in the mountains. Search out fields of wildflowers in the spring. Go hunting, fishing or birding, defying the definition of solitude as you are alone in the company of other species. Tend a garden or keep an aquarium welcoming other species onto your home turf. Allow yourself time to simply, quietly and closely observe other species, to see them for the amazing results of evolution that they are. Admire the structures that make them superbly suited to the life they live, which makes each unique among millions of others.

We can narrow the nature gap, too, by ceasing to take earth for granted. There are an estimated 10^{10} galaxies and 10^{21} stars. Among stars, perhaps one in five is orbited by planets. In spite of this mindboggling number of worlds, earth remains the only inhabited planet we know of, or are likely to experience. All the others are too far away, barren or both. We should be grateful that we find ourselves on this planet with its profusion of life, rather than some hostile corner of the Universe. We should appreciate the millions of life forms that challenge our intellects to understand, bring beauty into our lives and offer up a bounty of resources. We can resolve to close what is, in the big scheme of things, a recent gap between humans and nature, or risk a gulf becoming so wide that it can't be bridged.

Given human nature, I doubt that science can deliver society to a good place on its own. Yes, there are many aspects of mass extinction, conservation and environmental problems that must be approached with the cold, calculated rigor of disciplined thinking. But the appreciation of the natural world; a genuine love and respect for other species; a strong emotional connection to wild things and places requires us to be in touch with aspects of our humanity other than intellect. Robin Kimmerer's *Braiding Sweetgrass* is an example of a comfortable, effortless state of simply being that is pretty much the opposite of the noisy chaos of our lives today. Whether we as a society succeed in closing the nature gap depends on each of us finding our own path back to nature. For most of us, meaningful, enduring connections will not be found through carefully designed experiments, but in daring to break down the wall we have built between science and the best of our humanity. Science does not ask for or require compassion, love or altruism, but a bright future for humans and other species does. This does not mean diluting science in the slightest. Only allowing our motives, resolve and rewards to be enriched and bolstered by embracing parts of our humanness that lie outside science.

I have a particular cause in this book, to save the traditions of systematics and assure that we do not miss the opportunity to inventory, describe and classify species. This is necessarily associated with rebuilding *scientific* natural history alongside taxonomy. Careful, precise observations of organisms are an important adjunct to the exploration and classification of species, and important for learning to live sustainably. At the same time, I agree with Richard Louv that a wider cultural change is called for:

> We need, and I believe we see already growing, a cultural movement—what I call a New Nature Movement—that includes but goes beyond great programs

> that directly connect kids to nature: a movement that includes but goes beyond traditional environmentalism and sustainability, a movement that can touch every part of our society. The object is to give children the gifts of nature they deserve, and for all of us to find kinship with the lives around us, and the wholeness in the lives we live.

Few books have enjoyed such longevity as Anna Botsford Comstock's *Handbook of Nature Study*. Sitting on the science advisory board of Cornell University Press for many years, I was amazed by the enduring interest in her book. I came to realize that one sustaining constituency was homeschoolers, particularly religious fundamentalists. Not only did the book offer a great variety of accessible exercises observing nature, it did so with little mention of evolution, and it acknowledged the deity: "Out of this, God's beautiful world, there is everything waiting to heal lacerated nerves, to strengthen tired muscles, to please and content the soul that is torn to shreds with duty and care." Although I find evolutionary context to be among the most intellectually satisfying aspects of nature study, Comstock's approach nonetheless goes a good distance toward closing the nature gap by means of hands-on exercises:

> But, more than all, nature-study gives the child a sense of companionship with life out-of-doors and an abiding love of nature .... And these paths, whether they lead among the lowliest plants, or whether to the stars, finally converge and bring the wonderer to that serene peace and hopeful faith that is the sure inheritance of all those who realize fully that they are working units of this wonderful universe.

Whatever the path we find back to nature, whichever lens we choose to observe it through, getting to know other species, being able to identify them and appreciate details of their morphology and lives, are the keys to making biodiversity approachable and personally valued.

Further Reading

Bailey, L. H. (1915) *The Holy Earth*, Charles Scribner's Sons, New York, 171 pp.

Barlow, C. C. (1997) *Green Space, Green Time: The Way of Science*, Springer, New York, 329 pp.

Browning, W. D. and Ryan, C. O. (2020) *Nature Inside: A Biophilic Design Guide*, RIBA Publishing, London, 288 pp.

Comstock, A. B. (1990) *Handbook of Nature Study*, Cornell University Press, Ithaca, 887 pp.

Kimmerer, R. W. (2013) *Braiding Sweetgrass: Indigenous Wisdom, Scientific Knowledge, and the Teachings of Plants*, Milkweed Editions, Minneapolis, 408 pp.

Knapp, S. and Wheeler, Q. eds. (2009) *Letters to Linnaeus*, Linnean Society of London, London, 324 pp.

Louv, R. (1996) *The Web of Life: Weaving the Values that Sustain Us*, Conari Press, Berkeley, 256 pp.

Louv, R. (2008) *Last Child in the Woods: Saving Our Children from Nature-Deficit Disorder*, Algonquin Books, Chapel Hill, 323 pp.
Louv, R. (2012) *The Nature Principle: Reconnecting with Life in a Virtual Age*, Algonquin Books, Chapel Hill, 317 pp.
Louv, R. (2016) *Vitamin N: The Essential Guide to a Nature-Rich Life*, Algonquin Books, Chapel Hill, 277 pp.
Muir, J. (1877) Notes from Utah, part 4. *Daily Evening Bulletin*, 19 July, San Francisco.
Robertson, D. R. and Smith-Vaniz, W. F. (2008) Rotenone: An essential but demonized tool for assessing marine fish diversity. *BioScience*, 58: 165–170.
Schama, S. (1995) *Landscape and Memory*, Knopf, New York, 652 pp.
Suttie, J. (2016) How to protect kids from Nature-Deficit Disorder: Richard Louv explains how parents, educators, and urban planners can help kids reconnect with nature—before it is too late. *Greater Good Magazine*, 15 September.
Wilson, E. O. (1984) *Biophilia*, Harvard University Press, Cambridge, 176 pp.

17 Options for a Sustainable Future

Is it possible to create a future in which both humans and nature prosper? We had better hope so, because the path we are on is literally unsustainable. Unless we make dramatic changes, and soon, we will see the loss of millions of species. And risk, sooner than later, suffering a great reduction in the quality of our lives. We are too wasteful, too inefficient, too dependent on non-renewable resources and too complacent with the ticking environmental time bomb on which we sit.

Fortunately, it is not too late to change our ways. We can chart the biosphere, pole to pole, documenting all of earth's amazing life forms. If we do it right, paying attention to what makes each species distinct, we can open access to a wealth of clues with which to retool society. Tapping into millions of such tip-offs, we can dare dream of keeping earth species-rich while continuing to improve our quality of life through innovation. Importantly, as Janine Benyus said, "Unlike the Industrial Revolution, the Biomimicry Revolution introduces an era based not on what we can extract from nature, but on what we can learn from her."

Since the birth of humankind, we have turned to nature to solve problems of survival. Each time, we have been rewarded with ideas, materials or processes with which to overcome challenges. It's time to do so again. But this time on a grand scale, with intentionality and eagerness to learn. We can mine the diversity of life, striking paydirt time and again, on a mission to create options. Best of all, biodiversity is no normal mine. Not only is there an abundance of gems, but there are also no tailings. Only more possibilities we are not yet smart enough to recognize. As new problems arise, we will return to this groundmass, time and again, and rarely be disappointed.

We have entered a period unprecedented in human history. As the population grows, so too do environmental problems. Like promiscuous fruit flies in a bottle, the supply of bananas only goes so far. Given finite resources and space, there is an upper limit to how many humans our planet can support. Ehrlich's *The Population Bomb* may have not exploded on schedule, but the ecological principles he warned us about are as inflexible as any law of physics. Basic ecology and common sense tell us that the biosphere has a finite carrying capacity. Its sum primary production

DOI: 10.4324/9781003389071-19

is only enough to feed so many people. As Peter Raven has observed, if everyone in the world consumed at the level of people in the U.S., we would have already exceeded that capacity.

Non-renewable resources can be exploited only so long before they are exhausted. Landfills and hundreds of billions of pounds of plastics in the oceans are shameful evidence that we have not learned to recycle on the scale at which we cast off materials. The more people there are, the greater the impacts of fuel emissions, resource use, conversion of land, pollution and so forth. Seen from space, the earth we perceived to be fantastically large for countless generations suddenly looks small and vulnerable. Our current way of life simply cannot continue and change will be less painful now than waiting until things are even worse. It is essential that we find ways to lessen our footprint on the globe, to be more efficient and less wasteful in our industries, to give back to the world as freely as we take.

Some environmental problems can be avoided and should be. But many are here already, or so clearly visible on the horizon as to be inevitable. In those cases, we must learn to minimize their damage, alter our practices and protect the natural world. We must rethink the possible and overhaul our economy. And, if we wish to continue to improve the quality of our lives at the same time, we must innovate.

If history is prelude, there is reason for optimism. The march of civilization has relied on ingenuity and technology to avoid, overcome, mitigate and adapt—and we can continue to do so. Even E. O. Wilson, who saw threats facing biodiversity as clearly as anyone, ended *Half-Earth* on a hopeful technological note. My worry is that environmental challenges are increasing in number, scale and complexity faster than we are innovating. So, how can we give ourselves an edge in this race to solve problems before they overtake us?

In the past, many challenges were met by a combination of luck and dogged tenacity. Given the number of problems we face, it is unnerving to remain dependent on good fortune and protracted trial-and-error tinkering. To succeed in making a sustainable future, we will need inventors and entrepreneurs who refuse to give up. And we will need to arm them with hints that point to promising avenues for answers. Millions of such clues exist among species.

Sustainability, like most buzz words, has been used by so many people with varied agendas that it has lost precise meaning. Generally, it refers to the continuity of things we depend on, from natural resources to food production and development. Let me be specific. My vision of sustainability is a tomorrow in which humans dream and prosper in a self-perpetuating biosphere comprised of numerous and diverse species. This means, beginning today, we resolve to meet our needs without mortgaging prospects for future generations. It means taking care to avoid depletion of resources and degradation of ecosystems; allowing the biosphere to be expansive enough, and sufficiently species diverse, so as to be resilient; and working to conserve as many, and as diverse, species as possible. It is easy to enjoy the fruits of the biosphere today, leaving those who follow to fend for themselves. It is also immoral, if we can meet our needs while giving biodiversity and civilization their best chances tomorrow.

Biomimicry has emerged as the shortest, surest path to the creation of sustainable ways to design, build and do. Among species are innumerable solutions to the problems we face, already perfected, just waiting to be recognized and emulated. Janine Benyus makes clear by example the possibilities for biomimicry. By extension, it is evident that we have only begun to explore those possibilities. A couple of examples illustrate the breadth of problems that can be solved by biomimicry.

Getting a flu shot is an unpleasant and painful experience for many people who do not even notice when they have been bitten by a mosquito, until it's over and the itching sets in. This is because the mouthparts and saliva of the mosquito are precisely designed for sneak attack. Whether she is detected and swatted is a matter of life and death for a female mosquito in search of a blood meal. And without a blood meal, she cannot produce viable eggs. Engineers have mimicked mosquito mouthparts and designed a needle that, using the same mechanical principles, is effectively painless. Now, if they could do something about that itch.

There are termites in Africa that cultivate a fungus that grows best at 87°F. Outside temperatures vary from more than 100°F during the day to as low as 35°F at night. Termites have solved this fungus culturing problem by constructing a system of vents, regulating convection currents and maintaining the mound's internal temperature in spite of ambient fluctuations. Architect Mick Pearce and engineering company Arup mimicked termite nest design to engineer the Eastgate Centre in Zimbabwe, a building that cools itself using 90% less energy than conventional air conditioning.

From Velcro inspired by annoying burrs to Olympic swimsuits mimicking a shark's hydrodynamic skin, self-cleaning paints with microscopic surface properties of lotus leaves, and drones flying like hummingbirds, the possibilities in biomimicry are virtually endless. Natural selection has finely tuned species to solve problems of survival that we now or will soon face. Unlike inventors, natural selection never rests. It is a 24/7 research and development department that has, over millions and millions of years, given us multiple solutions for nearly every problem imaginable—including many we have yet to think of.

A major obstacle to nature-inspired solutions is the lack of information flow. Systematists describe attributes of species using highly specialized jargon largely indecipherable to engineers, inventors and entrepreneurs who might otherwise use them to spark their imaginations. Such terminology is necessary for scientific precision, but it need not remain an obstacle. Working with information scientists, we can crack this linguistic code, maintaining the jargon needed for taxonomic purposes while making evolutionary innovations accessible to non-specialists. Taxonomy opens access to the low-hanging fruit of biomimetics: untold millions of successful models for sustainable alternatives. We need only translate descriptions into non-technical language.

This means returning an emphasis to descriptions of morphology and scientific natural history observations. No amount of DNA can uncover the clues we need to fuel Benyus' biomimicry revolution. The search for biomimetic models will fundamentally change our perception of natural history collections, too. Far from warehouses of dead specimens, they will be seen as fountainheads of biomimetic creativity. For species that go extinct, museums will make it possible to continue to study and learn from them.

Having survived an aggressive form of prostate cancer myself, I am among those appreciative of efforts to find new and better treatments. The National Cancer Institute screened about 30,000 samples from diverse plants and animals between 1960 and 1981, in search of potentially useful chemical compounds. Paydirt was struck with taxol, discovered in the Pacific yew, *Taxus brevifolia*, and later synthesized. Because taxonomists group related species together in classifications, it was subsequently, and not surprisingly, found that other species of yew contain the compound, too. The Florida yew, *Taxus floridana*, is a case in point. Like the Pacific yew, the compound taxol is produced by an associated fungus and sequestered in the bark of the tree.

Imagine that taxol had been discovered in the Florida species instead and that it was the only living species of *Taxus*. Who would have imagined that an obscure tree, among the rarest in the world, could be the source of a treatment for breast and ovarian cancer? Restricted to wooded habitat along a 15-mile stretch of the Apalachicola River, this species is suddenly seen in a new light.

Such stories, within and far beyond the boundaries of biomedicine, will be repeated thousands of times in the years ahead if we follow through with a revival of descriptive taxonomy and an all-out species inventory. Some successes will solve simple problems, others will be transformational, spinning off entire new industries and solving what had seemed intractable problems. A plaque affixed to a boulder now marks the spot in the Gifford Pinchot National Forest where Pacific yew bark was first gathered. I am all for commemorating biomimetic success stories, but if we both support descriptive taxonomy and persist in erecting monuments, we shall soon be tripping over them everywhere.

Phylogenetic classifications make the search for similar, possibly better, biomimetic models among related species easy. But the same or similar solutions to a problem are sometimes converged upon by evolution in entirely unrelated species. A challenge for information scientists, then, is to develop search strategies for analogous attributes of species. This is not an impossible problem, but it may require changes in the practices of taxonomists to enable databases to be maximally informative. Refined search strategies should aim to not only locate a biomimetic clue when needed, but to find the very best one.

In order to accelerate solutions to the problems we face, we can do no better than invest in descriptive taxonomy and enabling software such as improved search engines. Species represent a mostly untapped, nearly bottomless well of biomimetic ideas. And this is the ideal time to connect taxonomy and biomimicry. An inventory will necessitate revisions and monographs that revisit named species as well as describing new ones. This is a superb opportunity to extract biomimetic clues from both. It need not require a separate initiative to liberate information, only an adjustment to descriptive taxonomy. Recognition of the potential of biomimicry could not have come at a more opportune time.

Further Reading

Allen, J. D. (2016) The proof is in the physics: Olympic 'sharkskin' swimsuits outperform shark skin, www.wshu.org/news/2016-06-28/the-proof-is-in-the-physics-olympic-sharks kin-swimsuits-outperform-shark-skin (Accessed 10 January 2023).

Caradonna, J. L. (2014) *Sustainability: A History*, Oxford University Press, Oxford, 344 pp.

Cohen, S., Eimicke, W., and Miller, A. (2015) *Sustainability Policy*, John Wiley & Sons, Hoboken, 272 pp.

Ehrlich, P. (1968) *The Population Bomb*, Ballentine, New York, 201 pp.

Goel, A. K., McAdams, D. A., and Stone, R. B. (2014) *Biologically Inspired Design: Computational Methods and Tools*, Springer, London, 325 pp.

Henningan, W. J. (2011) It's a bird! It's a spy! It's both. *Lost Angeles Times*, 17 February.

Izumi, H., Suzuki, M., Aoyagi, S., and Kanzaki, T. (2011) Realistic imitation of mosquito's proboscis: Electrochemically etched sharp and jagged needles and their cooperative inserting motion. *Sensors and Actuators*, A, *Physical*, 165: 115–123.

Liu, K. and Jiang, L. (2012) Bio-inspired self-cleaning surfaces. *Annual Review of Materials Research*, 42: 231–263.

Pfister, T., Schweighofer, M., and Reichel, A. (2016) *Sustainability*, Routledge, London, 138 pp.

Turner, J. S. (2010) *The Tinkerer's Accomplice: How Design Emerges from Life Itself*, Harvard University Press, Cambridge, 304 pp.

Vincent, J. F. V. and Mann, D. L. (2002) Systematic technology transfer from biology to engineering. *Philosophical Transactions of the Royal Society*, A, *Mathematical, Physical and Engineering Sciences*, 360: 159–173.

Vogel, S. (1998) *Cat's Paws and Catapults: Mechanical Worlds of Nature and People*, W. W. Norton, New York, 382 pp.

Part III

Solutions

It is past time to reimagine systematics for the 21st century, and beyond. The core vision, mission and questions that have driven systematists for hundreds of years are more relevant today than ever. Our challenge is to reembrace and build upon them, taking advantage of the extraordinary progress made to date, and restating taxonomy's aims in clear, unapologetic terms. The traditional goals of systematics are timeless, ultimate questions whose scientific, intellectual and societal importance are undiminished. Preserving the best of taxonomy's past, while embracing modern theories, technologies and practices, we can accelerate species exploration with no loss of excellence.

It is time to integrate all sources of relevant data, and put to use every appropriate tool. Morphology, paleontology, embryology and molecular genetics all have a place at the table. It is time to reassert systematic biology as an independent, fundamental, curiosity-driven science. To make clear that the purpose of systematics is to explore, discover, describe, name and classify earth's species phylogenetically—and those of other worlds, when they are discovered. Understanding species and characters themselves, in their historical context, is the aim of systematics and the beginning for countless applications of knowledge created. We can no longer permit taxonomy to be redefined in the image of some other science or supported for the sole purpose of meeting the needs of other biologists for species identifications. That is an essential application for taxonomic knowledge, but to make differentiation and identification of species our mission is short-sighted and self-defeating. Redefining systematics, deterring it from its own goals, will not only limit what we ultimately know and understand about species, characters and their history, it will reduce the reliability and information content of classifications and names.

So, keeping the time-honored aims of taxonomy intact, we must seek ways to accelerate its work in light of the extinction crisis underway. Modernizing taxonomy demands a makeover with three components: (1) taking advantage of the latest technologies to accelerate species exploration and taxonomic research, removing obstacles that separate taxon experts from research resources and colleagues; (2) expanding the workforce to include not only doctoral-level taxonomists, but also a cadre of specialized support staff and a trained, inspired army of competent citizen scientists; and (3) tackling the difficult challenge of balancing efficiencies of teamwork with the passion, creativity and intellectual freedom of individual species explorers.

DOI: 10.4324/9781003389071-20

18 Taxonomic Renaissance

What should taxonomy, reconceived to deal with a mass extinction, look like? Clearly, it should be ambitious. It may well be the last chance to discover millions of species. And efficient, time is of the essence. Increasing the yearly rate of species discovery by an order of magnitude seems a realistic initial milestone. It should insist on uncompromising excellence, assuring that we make the most of the opportunity that remains. With no do-overs, the temptation to take shortcuts must be avoided.

As Robert May observed,

> It says a lot about intellectual fashions, and about our values, that we have a computerized catalogue entry, along with many details, for each of several million books in the Library of Congress but no such catalogue for the living species we share our world with.

As extinction proceeds, it is increasingly critical that such a catalogue—and as many details about species as possible—exists, alongside comprehensive natural history collections.

Reminded of the central role of collections in systematics, museums should play a starring role as centers for taxonomic research and the preservation of specimens and data reflecting species diversity as fully as possible. Their collective responsibility in this crisis is daunting, to assure that every species is represented in one or more collections where it may be studied regardless of its status in the wild.

Systematics must be charged with gathering and integrating as much relevant evidence as possible from morphology to molecules and from fossils to developing embryos. Partnering with information scientists it should create a comprehensive biodiversity knowledge-base that can be mined for geographic, natural history, morphological, molecular and other information about species and higher taxa. This will function as a clearing house for all that is known of species for the benefit of anyone who can use it, from ecologists to conservation biologists and biomimicry problem-solvers.

Taxonomy should be unwavering in its mission to learn the combination of attributes unique to each species, to make names unique and reliable and to assure

DOI: 10.4324/9781003389071-21

classifications reflect up-to-date ideas about phylogenetic relationships. It should provide the means to accurately identify species as easily as possible, without straying from or compromising its own aims.

Its practice should include cyber-enabled taxonomy, or cybertaxonomy, a fusion of the traditional goals of taxonomy with cyberinfrastructure and digital instrumentation. It should assure that species and higher taxa remain scientific hypotheses, repeatedly tested in revisions and monographs to remain accurate reflections of our best understanding of species and groups of species. It should put a face on the biodiversity crisis by bridging the gap between people and nature, heightening awareness and concern for the plight of other species by making them known and accessible.

It should assist ecologists by creating baseline knowledge of species comprising ecosystems against which changes in species composition can be detected, monitored and possibly restored. It should be the world's Rosetta stone for biomimicry, translating attributes of species into accessible information. Its classifications should be a constant reminder of the evolutionary-historical connections among species. Opening biodiversity to observation at the granularity of species should enrich our lives. Heightening appreciation for species should call forth the best in our humanity and nurture a conservation *ethos*. Its public outreach should awaken our innate curiosity to learn about the origins and history of biodiversity.

A successful revival of systematics and worldwide inventory should demand an unprecedented degree of international collaboration and planning. Requiring participation by every willing nation, it should be woven as a tapestry from threads as diverse as our world's ecosystems and cultures. Although coordination is necessary, it should reflect the unique strengths and interests of participating countries. Some countries have retained expertise in descriptive taxonomy, others not so much. Some have invested heavily in DNA sequencing, others less so. Some have relatively limited faunas and floras, others are hotspots with massive numbers of undescribed species. Balanced smartly, all of these can be assets in a global campaign to explore species.

Guided by the belief that knowledge of biodiversity belongs to humankind, and should be grounded in physical evidence, natural history museums should serve as flagships for an armada of universities, institutes and government agencies. And it should humbly acknowledge that no combination of technologies, no level of describing, imaging or sequencing can substitute for the information content and value to science and society of well-preserved specimens.

A renaissance in taxonomy has two essential ingredients: a clear-throated reassertion of the mission and goals of taxonomy and modernization of infrastructure to facilitate the speed and efficiency with which systematists can achieve those goals. Rather than presenting a laundry list of what is needed to realize the modernization of taxonomy, from how we access and use museum collections to state-of-the-art instrumentation and cyberinfrastructure, I present a few fictional vignettes from one possible future for taxonomy. But first, let's clarify the vision for a revitalized taxonomy:

> The mission of taxonomy is to explore, discover, describe, name and classify every kind of plant, animal and microbe on earth; to reconstruct phylogeny (the pattern of evolutionary-historical relationships, at and above the level of species), expressing it in formal classifications, names, and cladograms (branching, treelike depictions); to do so by gathering and integrating every kind of relevant evidence from morphology to fossils, embryonic development and molecules; to foster scientific natural history observations in addition to making natural history collections a comprehensive representation of species diversity; to facilitate species identification and measurable conservation goals; and to unleash clues found among species to fuel a biomimetic revolution that leads us expeditiously into a sustainable future.

While I am justifiably frustrated with the state of taxonomy generally, it is important to state emphatically that there are many systematists working tirelessly to explore and make known species and to improve the lot of systematics. This is evidenced in thousands of new species discoveries each year, scholarly revisions and monographs (e.g., Wood et al.'s foundation monograph of *Ipomoea*), dedicated museum curators and continued advances in taxonomic theory and practice. I admire and applaud the excellent contributions from existing taxon experts, but the times demand much more for and from systematics.

For a glimpse of the systematics of tomorrow, let's time travel. Not far, just a few years from now. Just far enough into the future to see what a taxonomic renaissance might look like given proper funding, existing technology and a mandate to explore and inventory life on earth. If we fail to reimagine taxonomy then, like a Dickens' story, the future could be a very different one. A darker one. A future in which systematics is reduced to an identification service and limited to one source of data. A future where species go extinct randomly by the millions and society suffers for lack of options with which to adapt to the rigors of life on a rapidly changing planet. But, as Scrooge was told, such a bleak tomorrow is not the future that will be, only the future that may come to pass if we fail to mend our ways. Our visit will be to a brighter tomorrow. One in which we have returned to our senses, recognized the preciousness of biodiversity, resolved to support taxonomy, renewed scientific natural history and embarked on a worldwide campaign to learn all species. A future filled with the wonder of discovery and hope in conservation, and with a wealth of prospects for civilization, science and biodiversity.

With apologies for any confusion, I offer a word of caution. In the fictional account below, some things are mentioned that really do exist. For example, ZooBank has been created as a central registry for newly proposed scientific names of animals. In my imaginary peek into the future, I have enhanced ZooBank, introducing mandatory registration and automatic credits assigned to taxonomists each time a name is used. To avoid confusing fact with fiction, assume the story is made-up unless you consult authoritative sources that indicate otherwise. After all, science fiction is supposed to incorporate enough reality to be believable. And, what good is fiction if you can't make the world the way you would like it to be?

Cybertaxonomy

Dr. Alfreda Wallace is a world authority on fungus gnats and their relatives. There is a good chance you have been in the company of fungus gnats at one time or another, whether you knew it or not. Perhaps on a walk through the woods at dusk, when hordes of tiny insects take flight, just above ankle height, forming a sparse living cloud of aerial plankton. Fungus gnats are small in size, delicate in form and, to those of us who have a soft spot for the six-legged, elegant in an insect kind of way. The common name fungus gnat applies to about 3,000 species in the family Mycetophilidae, whose name is derived from word roots meaning, appropriately, "fungus loving," and another 1,000 species in the family Keroplatidae.

Alfie arrives at her lab early, espresso in hand, in a heightened sense of anticipation. Before leaving the night before, she had examined a museum specimen from South America that belongs to one of her favorite genera. It seemed to her almost certainly a new species. She settles into her work station in a laboratory more reminiscent of a fighter jet cockpit than a museum backroom.

To her left are state-of-the-art dissecting and compound microscopes. To her right, drawers of museum specimens, with row upon row of pinned flies, open for business. In front of her, a semi-circular wall of high-resolution digital displays, putting at her fingertips all the resources she requires for taxonomic decision-making: video conference links to colleagues around the world; texts and images from historic literature; an archive of digitized images of type (name-bearing) specimens; connections to remotely operable digital microscopes, allowing her to examine specimens, in real time, in distant museums and remote field sites; and, of course, a sophisticated assemblage of software specially designed to facilitate her descriptive and analytical work.

Collecting by Proxy

"Good morning, Alfie."
"Well?"
Silence.
"Come on Ray … you're killing me!"
"We got 'em!"
"Fantastic!" Ray leads a team of three collectors currently collaborating with Brazilian entomologists in the Atlantic coastal forest where a new species, *Neoceroplatus betaryiensis*, with bioluminescent larvae, was discovered in 2019 (unlike Ray, this tidbit is real). He and his team have trained with Wallace for several years to develop a sophisticated eye for the microhabitats where fungus gnats are likely to be found. While some species are collected in traps indiscriminately capturing flying insects, his team is uniquely adept at ferreting out rare and unusual species, as well as associating them with fungus hosts, or prey in the case of carnivorous species.

"I described the habitat conditions I was looking for, and our hosts, with their incredible knowledge of the forest, took us to just the right spot. Within an hour, we found an active colony in a branch on the ground. The whole thing, web intact, is on its way to you and should arrive this afternoon. We were able to collect about a dozen adults and estimate we have at least twice that number of larvae. Should be enough to get your rearing program going."

"I can't wait. Did you keep an adult that I can see now?"

"Sure did. I knew you'd ask, so I mounted one and held it back. Give me a second and you can see for yourself." The gnat has been mounted on a pin that Ray gently places under a portable, remotely controllable digital microscope. Alfie spins the insect around and increases the magnification to get a better view of the veins in the wing, then details of the legs.

"No question, Ray. That pattern on the wing really sets it apart from *betaryiensis*. It's the new *Neoceroplatus* ... Good work!"

Wallace is collaborating with other taxonomists, and the research and development team of a leading LED manufacturer. They are comparing bioluminescence in several species of flies, beetles and deep-sea fishes, with the hope of developing new, chemical-based, lighting technology for use in emergency and remote situations where electricity is unavailable. This biomimetic project is one of several spin-offs of her research, such as an investigation of the physical properties of silk spun by fungus gnat larvae for a textile manufacturer. Her personal interests include the evolution of bioluminescence. Molecular analyses suggest that this is one of the several closely related species that emit a blue light, but among them is at least one non-bioluminescent species. Was bioluminescence lost in this species? Or, were the genetic precursors present in an earlier ancestor and luminescence evolved in parallel two, or more, times among species of the genus? Beyond producing larvae for the bioluminescence project, and silk for a biochemist to analyze, she hopes to corroborate tentative evidence that larvae of this new species feed on fungi, rather than being predatory. The use of such divergent food sources is another interesting angle on fungus gnat evolution. She relies on two teams of professional collectors to gather material for her species inventory and function as Specialized Worldwide Advanced Taxonomic (SWAT) teams to help fund her lab.

Send in the SWAT Team

The telephone rings. It's the director of an ATBI in South Africa. All-Taxon Biodiversity Inventories are projects that aim to account for every species that lives at an ecological study site. No number of ATBIs around the world can meet the needs of taxonomists, like Alfie, that are quite different from those of ecologists. The taxonomist wants to know all species descended from a common ancestor, such as all fungus gnats, regardless of where on earth they live. This taxon focus is distinct from a place-defined approach, like an ATBI. All the same, valuable specimens come from ATBIs, and it is an important service to ecology and conservation to contribute expertise to such local inventories. In this brave new world, ATBIs have emerged as one of several sites for biomimicry think tanks, along with taxonomic centers of excellence in museums and universities. It turns out that the immersion of industry representatives in wilderness inspires out-of-the-box thinking.

It makes sense to have ecological researchers and conservation biologists on site at an ATBI, but with so many kinds of organisms there is no way to support in-house taxonomic expertise for more than a few groups at most. Can you imagine one ATBI site, such as a national park, having experts on earthworms, mosses, fungi, tardigrades, birds, algae, protists and fungus gnats, among hundreds of others? Even if it were possible, it would be a waste of money. Far better to permit taxonomists to pursue a worldwide inventory for their own purposes, inviting them to contribute to local floras and faunas, make specialized expertise available and be on call for tricky identifications.

The ATBI director is phoning to request an inventory of fungus gnats at his site. To date, they have collected only half a dozen common species, but the fungal diversity is high and he suspects, based on numbers found in similar habitats elsewhere, there must be many more. He is pleased to learn that one of Wallace's SWAT teams has just returned from an assignment in Eastern Europe and can be deployed for a few months to carry out an intensive survey.

SWAT teams are a new idea to accelerate species discovery in the global inventory. They consist of extensively trained and experienced professional collectors who have deep, specialized knowledge of a taxonomic group (fungus gnats, in this case) including where they live, what hosts they use, how to collect them, how to preserve them for anatomical and molecular study, how to voucher host specimens and, importantly, how to identify common species on sight, so as not to waste time redundantly collecting the same species over and over. Their mission is to discover new species and increase knowledge of named ones. At ATBIs, they often work with staff to assemble a synoptic collection of local species. Their advanced training, and seat-of-the-pants experience, make SWAT teams effective at finding rare and new species. Further, SWAT teams complement data from ATBIs by collecting at many other locations where fungus gnats might live.

Teamwork

Taxonomy is both an individual and team sport. Alfie is a leading participant in an international network of fungus gnat taxonomists, students and volunteers. Before the inventory, there were few mycetophilid experts anywhere. A combination of educating young specialists and coordinating fungus gnat work around the world is accelerating the growth of knowledge of this ancient lineage of flies. Her team coordinates collecting, maintenance of a knowledge-base, growth and development of collections, advanced taxonomic education and training and responses to special needs for species identifications, providing a strong return on every dollar invested in her work. She has collectors in the field. Preparators pinning and labeling specimens. Sorters making tentative identifications and flagging unusual specimens. Technicians rearing species, dissecting specimens, preparing microscope slides and freezing tissues for DNA analysis. A mycologist identifies hosts and coordinates collecting of fungi with flies. Additional staff prepare drawings, take digital images, enter and manage data, take measurements, carry out literature searches and check the work of volunteers. Active taxonomists generally have at least three support staff, lessening the burden of routine tasks. For some top researchers, and centers of excellence, staff also includes curators, instructors, administrative assistants and public relations experts.

Individual researcher-curators and international teams, like Alfie's, appear at first to be conflicting models, yet it works. Taxonomy remains an intensely individual kind of research, with knowledge, hypotheses, classifications and species names associated with the person generating them. While decisions are taken by individual scientists, they are informed by consultation, debate and peer review within a vibrant, interactive community of researchers who share the vision of a complete inventory, tested and corroborated hypotheses about species and relationships and highly informative and reliable names and classifications.

Citizen Scientists and Crowd-Sourced Identifications

Wallace has developed a routine that puts citizen scientists to work. She has scores of them. From students to retirees and weekend bug warriors, they are fanned out across the globe making important contributions to natural history observations of fungus gnats, filling gaps in knowledge of geographic distributions and documenting host and prey associations.

She is especially proud of a dedicated group of about 50 volunteers who have completed rigorous, self-guided, online training to become qualified to identify common species of fungus gnats. As specimens are collected, pinned and labeled, she uses an automated imaging system to create a digital 3D representation of each unidentified specimen. These images can be accessed on the internet by qualified volunteers, spun around and zoomed in on, just like a museum specimen under a microscope. Because digital imaging is automated, it takes only minutes to create such composite images that act as surrogate specimens that can be examined from anywhere.

Only volunteers who have passed a strenuous examination, demonstrating their mastery of morphology and ability to identify species, are given access to the online backlog of "unknowns." When the majority of volunteers identify a specimen as the same species, they are almost invariably correct. But when they disagree, arriving at two or more identifications, the specimen deserves the attention of Dr. Wallace or another expert. Conflicting determinations may be due to mutants, or damaged or poorly mounted specimens. Or they may represent new species. This crowd-sourced provisional identification allows motivated amateurs to contribute in hands-on fashion to science, enjoy the intellectual satisfaction of consequential work and free up the time of professionals for other activities.

e-Monography

Dr. Wallace is currently working on a taxonomic revision of one of the largest genera of fungus gnats. Using specially designed software, she has all the information resources, instrumentation and contacts necessary to efficiently write descriptions, analyze relationships, consult with colleagues and confirm that names are used correctly for existing species. Rather than a printed monograph that would be out of date as quickly as the next species is discovered, she is constructing an electronic monograph for the family, genus by genus, that serves as a comprehensive taxon knowledge-base and a one-stop shopping site for everything known about Mycetophilidae. She is one of several specialists around the world who collectively maintain this community e-monograph. Her work, like that of other participants, is subject to rigorous peer review, and her contributions are acknowledged by embedded author and date metadata.

She has just discovered an unusual new species and completed a description of its morphology, accompanied by high-resolution images. The genome was sequenced by a lab she contracts to extract and analyze DNA. She submits her species description to a scientific journal editor for review. With the click of a submission button, she sets in motion a cascade of actions. An invitation is sent to qualified reviewers drawn from a database of experts. Each has instant access to her manuscript. Unlike the past, when peer assessment of descriptions was limited to her written words and whatever images she provided, reviewers can now access actual specimens by means of telemicroscopy and judge the accuracy of her work, and the significance of unusual morphology, for themselves. Within days, sometimes hours, her tentative species will have been critically reviewed and either endorsed for publication, or suggestions made to improve or abandon its description.

Once accepted, her new species is sent by a journal editor to ZooBank, a central registration site. It verifies that the name conforms to rules of the international code governing scientific names for animals, has not been previously used, and assures the editor that it has been spelled and used correctly throughout the manuscript. Upon publication, the name is added to a global catalogue and the existence of the new species is instantly made known to the world. Name changes and names of new species do not become official until they are registered. There is no censorship, only checking for spelling and compliance with rules. Software also keeps track of species names in all scientific publications, giving credit to the taxonomist who named a species each time the name is used. This has leveled the field for taxonomists who did not fare well in citation indexes in the past.

An additional requirement of registration is that digital images of diagnostic characters be deposited in an open-access archive, along with the text of the description. No copyright restrictions exist as an obstacle between taxonomist and users of taxonomic knowledge. Only analyses and discussions in taxonomic papers can be copyright restricted by journals. Descriptions, including images, are considered property of the scientific community. Major natural history museums have e-typification imaging centers that help assure as many diagnostic features as possible are visible in images.

Type specimens are actual, physical specimens deposited in museums to which scientific names are attached. They act as a kind of international standard for scientific names. As ideas about what a species is, and how many species there are, change over time, wherever the type specimen fits in the latest scheme of things, there follows the name attached to it. This effective system, with the objectivity of an actual specimen, strikes an ideal balance between stability and flexibility.

Imaging centers make a series of high-resolution digital images that collectively constitute an "e-type." This means that, for many routine purposes, users can simply consult a library of images, or a composite 3D image, and need not handle the actual specimen. On other occasions, when existing images are not enough, the type can be placed under a remotely operable microscope by a technician or curator for further observation and imaging by an expert. When both images and remote access prove insufficient, taxonomists visit museums for direct, hands-on study, such as when dissection or other manipulation is required. But even then, a few museums have modified da Vinci surgical robots that allow a specialist to dissect a high-value specimen from a remote location, with the assistance of a local staff member. As experts reexamine specimens over time, additional digital images and annotations make the e-type increasingly informative.

One or more museums in every country offer a free digitization service for those who deposit types in their collection. For others, the service is available for a nominal fee. When publication and registration make her new species official, her description is automatically translated into less technical form and a page is created in an open-access, comprehensive, encyclopedic source of species information online, along with links to other pertinent information. The e-monograph and geographic distribution maps are automatically updated, as are the default phylogeny, formal classification, regional checklists, diagnostic identification tools and online field guides.

Users of the e-monograph can generate publications and reports on demand, such as a checklist of species in a particular state, zip code or national park; a field guide with all and only those species found at an ecotourism location; or an accounting of all known genetic and geographic variations within a species. Ecologists find it convenient that identification keys can be tailored for their study sites. By excluding extralimital species, these keys are shorter and simpler to use than those that include all species worldwide. Dichotomous keys give the user alternative choices (e.g., are the wings clear, or do they have a color pattern?) and, like a game of 20 questions, soon lead to an identification. Alternatively, DNA barcodes may be chosen to confirm identities, and artificial-intelligence-based software can often identify species based on submitted images.

Just as generations of birders have maintained life lists, communities of enthusiasts now exist for all kinds of organisms, from diatoms to dragonflies, cycads to centipedes, similarly driven to see as many species as they can in a lifetime. Software is available that seamlessly combines identification tools with a running life list, automatically keeping pace with taxonomic changes. Such e-life lists can be searched and visualized in many ways, from a chronological account of sightings to a map of excursions. Ecotourism companies offer special international trips led by taxonomists, so that enthusiasts can see species among the rarest on earth and ask questions of a leading expert. These taxon discovery tours help build strong citizen support for unusual groups of species and are a supplemental source of income for scientists, institutions and local conservation projects.

e-Monographs assure that users have immediate access to the most complete, up-to-date species information possible. A species newly named is available to the world instantly. No longer limited by out-of-date, printed monographs, professional and citizen scientists have the opportunity to contribute observations about the geographic distribution and natural history of recently discovered species. Amateur observations are flagged for review, first by crowd-source assessment by peers, then review by a recognized authority who can assure its accuracy before it is integrated in the verified knowledge-base. Because metadata allows the user to dig deeper into the source of each record, it is possible to filter out amateur IDs, seeing only determinations made by trusted authorities.

There are legitimate reasons that experts may differ in their assumptions and interpretations regarding characters, species and classifications. After all, they are hypotheses. And hypotheses have background assumptions. While there is an explicit, default set of phylogeny assumptions built into each e-monograph, software permits users to do their own analysis on the fly—no pun intended—using their own assumptions, or call forth results of analyses that have been done by various experts.

The brilliance of e-monography goes beyond its up-to-the-minute information content. It is also a tool guarding intellectual freedom. While the community of fungus gnat experts has reached consensus opinions on many things, there are always disagreements over details and lone wolves who see things very differently. It is critical that minority hypotheses, whether of homology or classification, not be silenced by the majority, or worse, some all-powerful ruling committee. Throughout history new and sometimes radical ideas have been dismissed, only to be shown to be correct at a later date. The e-monograph makes it possible to present multiple, alternative species hypotheses, phylogenies, character interpretations and classifications for the benefit of the specialist researcher, alongside a consensus or default version, generally based on the most recent "published" monographic treatment, to keep things simple for general users.

Taxonomic stability is seen as a virtue by users of taxonomy but is a threat to science. For convenience, everyone would like to see names and classifications never change, but such stability comes at the expense of accuracy and the growth of knowledge. As taxonomic hypotheses are tested and specimens collected, ideas will change, and so too must names and classifications. Everyone committed to science should welcome improved knowledge and names and classifications that keep pace with it.

Doing Your Own Thing … Together

Diversity of evidence and increased standards in taxonomy have encouraged both specialization and cooperation in the interest of excellence and efficiency. This recognition has proven liberating for everyone. Whatever your preferences or skill set, it is an all-hands-on-deck effort and there is no shortage of every kind of work. Some taxonomists concentrate on one data source, others incorporate two or more in their work. All are needed, welcomed and encouraged. Progress is measured by collective advances of the taxonomic community, with sharing and synthesis of data the norm.

This has resulted in a flourishing of both specialization and collaboration. There are collectors who prefer life in the field, rarely sleeping under a solid roof. Morphologists focused on in-depth morphological studies. Molecular systematists busily sequencing genomes, populating a GenBank that is, with the resurgence of taxonomy, bursting with authoritatively identified species—species that represent carefully constructed, corroborated hypotheses. There are paleontologists digging, and embryologists studying developmental pathways by which tissues differentiate into diverse and complex characters. In the end, it all comes together. Like talented musicians in a well-rehearsed orchestra, an array of specializations, creativity and passions converge in a classification and e-monograph that is a crescendo of taxonomic excellence and understanding. In the case of fungus gnats, Dr. Wallace is the conductor.

Modern Museum

Accessing museum specimens is no longer an obstacle to taxonomy that requires experts to travel to many cities to complete a study. Specimens are stored on open shelving in sealed rooms. Pests and environmental fluctuations that once threatened collections are kept at bay by physical exclusion and closely maintained environments. The occasional breakage of rare, delicate specimens in the hands of well-meaning experts is virtually a thing of the past as robotic arms precisely retrieve specimens and place them under remotely operated digital microscopes.

As newly collected, unidentified specimens come into the museum they are prepared and mounted, 3D imaged, then placed on trays in a grid pattern and photographed in extremely high resolution. Trained volunteers scan these images online, zoom in on specimens, and tentatively sort them so that they are available to experts faster. Crowd-sourced identifications, like those Wallace uses in her research, weed out common species allowing specialists to efficiently use their time.

Public education programs and exhibitions proudly share species discoveries of staff members, teaching the public how modern systematics works and sharing the excitement of exploring life on our little-known planet. Museums have become leaders in systematics, species inventories and educating the public about the diversity, status and history of life. Visiting a natural history museum you enter another world, the world of taxonomy, where our planet's species-scape is coming into focus one species at a time.

Biomimetic Ideation

Taxonomists, like Dr. Wallace, embrace the opportunity to partner with private industry, exploring biomimicry as a path to a sustainable future. They recognize that their discoveries, descriptions, names and classifications are the portal through which a new generation of designers, engineers and entrepreneurs will find the inspiration for world-changing innovations. Centers of taxonomic excellence are connected to biomimetic think tanks. Inventors and industry scientists come to such centers with problems in search of solutions, finding a creative, welcoming environment in which to mine species for promising leads.

Like all creative processes, biomimicry is not an exact art. Some of the best solutions require connecting facts that no one had ever thought to associate. This is something that the human mind is particularly good at, and which computer programs generally cannot accomplish with similar ease. How far artificial intelligence will progress in this respect is yet to be seen but, at least for now, bringing together individuals with different sets of knowledge is the synergistic grease on the axel of innovation.

Speaking of AI, taxonomists were early to recognize its promise. Norman Platnick of the American Museum of Natural History developed software years ago that "learned" the range of variation in the genitalia of spiders as more and more specimens were imaged (see Russell et al.). In the end, his software could identify species with more than 90% accuracy that, previously, had required an expert's eye. This work pioneered the use of AI for species identification, something becoming more commonplace in taxonomy, e.g., CSIRO's Critterpedia project.

My experience in libraries makes the power of unforeseen connections of information intuitively obvious to me. When I was an undergraduate student, my favorite place to study was a desolate aisle in deathly quiet stacks of Ohio State's Botany and Zoology Library. Being surrounded by scientific books close to my interests was a source of inspiration and comfort. And, when I took breaks from my studies, I would randomly browse the shelves. There were many times when scanning through a book I would never have thought to purposefully consult led me to make an unexpected connection of ideas. Many biomimetic breakthroughs will be like that, random, unplanned, lucky convergences of facts, needs, people and ideas.

Perception through Reality

In my imagined future, all is not yet rainbows and unicorns. Extinction, although slowed, continues. And systematics, while progressing rapidly, is still clawing back its reputation. Systematics was revitalized first and foremost by reclaiming its identity, remembering its core mission and goals. These traditional aims have been strengthened, and made efficient, by adopting recent advances in the information and digital sciences and popularized by trumpeting successes. Institutions have committed to systematics and the duration of the inventory whatever fads may come and go. Public pressure has catalyzed change. People concerned about extinction and loss of knowledge have lobbied their elected representatives demanding species exploration and systematics be top priorities. And politicians have listened.

As species exploration has ramped up, advances in conservation, astonishing species discoveries, revelations about evolutionary history and biomimetic breakthroughs have captured headlines and the public's imagination, letting the importance of systematics speak for itself. Perceptions are now following this new reality. If this is excessively optimistic, then it is so only by degree. Once the ball is rolling, it will fall upon taxonomists themselves to be tireless promoters of their ideas and accomplishments. Systematists allowed the marginalization of their science to happen by trying to appease colleagues, chase easy money and conform to fashions. Unless taxonomists find their spine and become forceful, unapologetic advocates for their science, no one else will. A new and accurate perception of systematics as an independent, leading-edge science engaged in the biological dimensions of planetary science is propagated by museums, universities and researchers. Success breeds success.

Returning to present realities, the urgent need for taxonomic knowledge should be motivation enough for change. This is a rare moment in history when the stars have aligned to enable the return of systematics to its own mission, elevated in importance and urgency by the extinction crisis. This may well be the last chance for systematics before the majority of species, and species experts, are gone. Only systematists can resist pressures to conform to fads, be advocates for their science, inspire the public with their historic mission, and mobilize their community to finish what Linnaeus started.

Further Reading

CSIRO (2020) Critterpedia: An AI-powered app to identify insect and snake species. *Imaging and Computer Vision*, 15 July. https://research.csiro.au/icv/critterpedia-an-ai-powered-app-to-identify-insect-and-snake-species/ (Accessed 10 January 2023).

Falaschi, R. L., Amaral, D. T., Santos, I., et al. (2019) *Neoceroplatus betaryiensis* nov. sp. (Diptera: Keroplatidae) is the first record of a bioluminescent fungus-gnat in South America. *Scientific Reports*, 9: 11291. doi:10.1038/s41598-019-47753-w

Grace, O. M., Pérez-Escobar, O. A., Lucas, E. J., et al. (2021) Botanical monography in the Anthropocene. *Trends in Plant Science*, 26: 433–441.

Knapp, S. (2008) Taxonomy as a team sport. In *The New Taxonomy* (ed. Q. Wheeler), CRC Press, Boca Raton, pp. 33–54.

Martin, B. (2020) Research grants and agenda shaping In *Groupthink in Science* (eds. D. M. Allen and J. W. Howell), Springer, Cham, pp. 77–84.

May, R. (1988) How many species are there on Earth? *Science*, 241: 1441–1449.

Mayo, S. J., Allkin, R., Baker, W., et al. (2008) Alpha e-taxonomy: Responses from the systematics community to the biodiversity crisis. *Kew Bulletin*, 63: 1–16.

Milius, S. (2008) Biological moon shot: Realizing the dream of a Web page for every living thing. *Science News*, 173: 72–73.

Muñoz-Rodríguez, P., Carruthers, T., Wood, J. R. I., et al. (2019) A taxonomic monograph of ipomoea integrated across phylogenetic scales. *Nature Plants*, 5: 1136–1144.

Padial, J. M., Miralles, A., De la Riva, I., and Vinces, M. (2010) The integrative future of taxonomy. *Frontiers in Zoology*, 7: 16. doi:10.1186/1742-9994-7-16

Pavlinov, I. Y. (2021) *Biological Systematics: History and Theory*, CRC Press, Boca Raton, 255 pp.

Reginato, M. 2016. monographaR: An R package to facilitate the production of plant taxonomic monographs. *Brittonia*, 68: 212–216.

Russell, K., Do, M. T., Huff, J. C., and Platnick, N. I. (2007) Introducing SPIDA-Web: Wavelets, neural networks and internet accessibility in an image-based automated identification system. In *Automated Taxon Identification in Systematics: Theory, Approaches, and Applications* (ed. N. MacLeod), Taylor and Francis, London, pp. 131–152.

Scoble, M. J. (2008) Networks and their role in e-taxonomy. In *The New Taxonomy* (ed. Q. Wheeler), CRC Press, Boca Raton, pp. 19–34.

Wheeler, Q. (2007) Invertebrate systematics or spineless taxonomy? *Zootaxa*, 1668: 11–18.

Wheeler, Q., ed. (2008) *The New Taxonomy*, CRC Press, Boca Raton, 237 pp.

Wheeler, Q. (2020) A taxonomic renaissance in three acts. *Megataxa*, 1: 4–8.

Wheeler, Q. (2020, May) An unfinished revolution: In response to "A Quiet Revolutionary." Letters to the Editors, *Inference*, 5, doi:10.37282/991819.20.29

Wheeler, Q. and Miller, K. B. (2017) The science of insect taxonomy: Prospects and needs. In *Insect Biodiversity: Science and Society*, 2nd ed., vol. 1 (eds. R. G. Foottit and P. H. Adler), John Wiley, Hoboken, pp. 499–526.

Wheeler, Q. and Valdecasas, A. G. (2007) Ten challenges to transform taxonomy. *Graellsia*, 61: 151–160.

Wood, J. R. I., Muñoz-Rodríguez, P., Williams, B. R. M., and Scotland, R. W. (2020) A foundation monograph of *Ipomoea* (Convolvulaceae) in the New World. *PhytoKeys*, 143: 1–823.

Yeates, D. (2009) Viva la revolución: Designing the digital renaissance in zoological taxonomy. *Australian Journal of Entomology*, 48: 189–193.

19 A Planetary-Scale Species Inventory

Is it possible to complete an all-species inventory before millions of species are gone? Thanks to generations of taxonomists and existing collections the answer is an emphatic, if qualified, yes. Qualified, because fewer species remain to be discovered each year making "all" a moving target. Acting with urgency, however, we can create a relatively complete picture of diversity as it exists.

Linnaeus pointed the way with his vision of discovering, naming and classifying every species. Since the 18th century a small, dedicated army of systematists, professional and amateur, have doggedly pursued this vision. Presently, we add fewer than 20,000 new species per year. If we wish to achieve a comprehensive inventory rapidly enough to document the diversity of life and inform conservation policy, then this pace must be dramatically increased.

Reversing the loss of taxonomic expertise will be time-consuming. Full-time, professional taxonomists, curricula and graduate programs have all suffered in recent decades making it difficult for aspiring students to be formally educated and properly mentored by seasoned experts. However, as the taxonomic education system is rebuilt, no time needs to be lost. By supporting existing taxonomists to devote full attention to species exploration, and providing them with a few support staff, their productivity could reasonably be increased to at least 100 species treatments per year per expert. This includes both new species described and existing species tested and redescribed. Modernizing systematics' infrastructure, transformational efficiencies are possible with current obstacles to progress reduced or eliminated. Organizing international teams of taxonomists and institutions could leverage existing expertise and resources to accelerate discovery, too. And, supporting students and postdocs to work with experts can assure that specialized knowledge is passed on to the next generation.

At a workshop in New York, I invited a group of scientists and computer engineers to examine the challenge of an all-species inventory. They concluded that a combination of more taxonomists, teamwork and infrastructure modernization could, within five years, increase the annual rate of species discovery or corroboration by at least an order of magnitude to 200,000 species per year. At that rate, 10 million species could be made known—be described, or redescribed, as necessary—in 50 years, possibly less.

DOI: 10.4324/9781003389071-22

A comprehensive, worldwide taxonomic inventory of species ranks among the most ambitious science projects ever conceived. Few initiatives match it in scale, potential impact or urgency. Time to carry out an inventory is disappearing fast, unless we resign to live with numerous, large and random gaps in our knowledge of the biosphere, biodiversity and evolutionary history.

The goal of an inventory is deceptively simple sounding: to discover, describe, name and classify all species. Because systematics involves deep scholarship, rigorous research, extensive hypothesis testing and intensive field work, the orchestration of an inventory is anything but simple. Fortunately, we begin with several things in our favor. In spite of waning support in recent decades, there remain taxonomic experts ready and eager to scale up their efforts. Advances in theory mean that hypotheses about species, characters and relationships are more rigorously scientific than ever before. Advances in technology can accelerate every phase of inventory and descriptive work, without sacrificing excellence. And, with 2 million species already described, a framework phylogeny and hundreds of millions of specimens awaiting study in natural history museums, we are off to a good start.

As a common-sense prelude to an inventory, we should do nothing stupid like throw away hundreds of years of progress. Two such ill-conceived proposals are afoot. First is the idea of a molecular-based taxonomy that would marginalize or ignore, rather than build upon and expand, the great achievements of comparative and descriptive morphology and paleontology. Much better that we continue the tradition of pursuing all relevant evidence, adding molecular data to the mix while abandoning none. Second is the so-called phylocode, discussed previously, that would replace the Linnaean system of ranks and names with a system of names that contain less information and lack the flexibility to adapt to the growth of knowledge or changing scientific paradigms. Both proposals have a Kardashian quality to them, being popular for being edgy without offering worthwhile advances. Can molecular methods be integrated into taxonomy? They already have been. Can the Linnaean system be tweaked to overcome any perceived weaknesses without starting from scratch? Of course, and mostly it has been.

It is important to make two things clear from the outset. First, the goal is to *know* species, including their unique and shared attributes, not merely learn that they exist; to understand them and their characters in an evolutionary-historical context, not simply tell species apart; and to make each a testable hypothesis. Second, descriptive taxonomy is not a one-time activity. A leading ecologist once opined that after species have been found, described and named, that taxonomy's job is done and we can turn our attention to ecological, behavioral and other kinds of studies. This is false and shows a complete lack of understanding of systematics. Unless taxonomy is made arbitrary and fixed, in which case it would no longer be science, it must remain an ongoing enterprise. Species exploration and natural classification is a process, not a one-time activity. Taxonomic hypotheses are repeatedly tested as new characters are discovered, specimens collected and relationships analyzed. Thus, an inventory with enduring scientific and societal benefits includes an intensive initial period of exploration and discovery followed by continuing systematics research that expands and assures the reliability of knowledge.

The urgency attached to a species inventory cannot be overstated. Millions of species not discovered in this century will never be known. Many species that could have been saved will go extinct unless we get a better handle on biodiversity and its distribution. Gaps will exist in the story of evolution in proportion to the number of species left undiscovered. Our ability to refine and improve classifications will suffer unless specimens and observations are collected now. We will stumble in our race toward a sustainable future if we fail to document as many attributes of as many species as possible, including those soon to be extinct.

It is not good enough to commit to an all-species inventory without a target completion date. Linnaeus is proof of that. Unless we challenge ourselves with an ambitious deadline, a large number of species will have gone extinct before a first-pass inventory is completed. President John F. Kennedy turned up the heat on NASA to accomplish a moonshot within ten years. We must do the same. With millions and millions of unknown species, finishing the inventory in a decade is impossible. Fifty years, however, is a realistic stretch goal. Suggested first by Peter Raven and Ed Wilson, then reaffirmed by a gathering of scientists and engineers, 50 years is both ambitious and possible.

So, where do we start? Two very different ideas have been suggested. The first recommends focusing on comparatively well-known taxa, groups for which we can build on an extensive foundation of knowledge to complete an inventory relatively quickly:

> Some systematists have urged the initiation of a global biodiversity survey, aimed at the ultimate full identification and biogeography of all species. Others, noting the shortage of personnel, funds, and above all, time, see the only realistic hope to lie in overall inventories of those groups that are relatively well known now, including flowering plants, vertebrates, butterflies, and a few others.
>
> (Raven and Wilson, 1992)

Because this idea was introduced by a botanist (Raven) and myrmecologist (Wilson), it has been affectionately dubbed the "plants and ants" plan. Given the current situation, there is much to commend it. For such groups, we are well on the way to a comprehensive inventory and have a good idea of where to look for species not yet discovered. Completing such focused inventories offers proof of concept, giving credibility to the idea of the inventory as a whole. And being well known, such "popular" groups tend to have established specialists to immediately take up the work.

Raven's view is shaped by a recognition of the frightening speed with which biodiversity is disappearing, and is both precise and practical:

> my view is that what we ought to do is inventory aggressively for the groups that are well enough known [to] get as complete as possible a picture of their

> distributions while we still can—terrestrial vertebrates, plants, butterflies, mosquitoes, some groups of beetles, and a few others—and that for the rest, like nematodes, mites, and fungi ... we try to figure out sampling procedures (a comprehensive sample every hundred miles, or whatever, in a worldwide grid, for example) so that we could gain some idea of their distribution patterns while they are still there.
>
> (Personal correspondence)

No one has a better grasp of what's happening around the world than Raven, but I am not willing to write off an all-species inventory just yet. Even if it proves futile in the fullness of time—and I believe it will not, if we act decisively and soon to restore systematics—I would rather go down fighting than surrender entire taxa, permitting them to be forever poorly known. We can revitalize taxonomy and increase knowledge of species in every group, even if differing degrees of completeness are ultimately realized. My goals go beyond a picture of the ecological and geographic distributions of species. I am even more concerned with filling in details of evolutionary history and the diverse attributes of species.

I see equally compelling merit in a different approach. Pretty much the opposite strategy, it would give priority to least well-known groups. It is more of a bugs-and-slugs or ferns-and-worms plan, focused on hyper-diverse and neglected taxa for which we know only a small fraction of species. Consider the fungi. Only about 70,000 species have been named. Recent molecular work suggests that fungi could outnumber plants by a ratio of as much as six to one. If this is so, then the total number of fungi could be more than 5 million species! Fungi are not alone, with similarly startling examples among invertebrate animals, both terrestrial and marine. Round worms make the point. There are about 28,000 named nematodes, while experts believe the total number is several hundred thousand to as many as 1 million species. The argument for starting with least-known groups is that we can realize greater immediate returns on investment—at least as measured in numbers of species. For such taxa, it is difficult to carry out expeditions to the right kinds of places and not come back with buckets of new species. This gives a different sense of progress with lots of species discovered in rapid-fire succession.

Common sense suggests that we pursue a mixture of the two. Pick some taxa, like flowering plants and ants, and resolve to complete an inventory as quickly as possible. At the same time, select some hyper-diverse groups, such as mites or parasitic wasps, where discoveries will flow like water. It would be nice if all taxa could progress equally rapidly, without bias or favor, but numbers of species, and limited resources and expertise, will dictate that progress be made at different rates in different taxa. And, sadly, it may mean, too, that different taxa end up known to differing degrees of completeness.

Ask ten taxonomists which groups should come first and you will hear as many different answers. We should learn from astronomers and astrophysicists to make difficult decisions about sequence and priority with robust community participation, but outside the public eye where squabbles make the community appear disorganized. Confidence in the enterprise is enhanced by visible community-wide support. By creating a rolling series of decadal plans, astronomers have managed to focus on their long-term vision while achieving impressive milestones along the way. Regardless of how the community prioritizes species exploration, replenishing expertise must proceed alongside. Because students will gravitate toward varied groups, education should be two-tiered. Basic courses in theory and methods should create a common curriculum of excellence followed by training and apprenticeships for developing specialized knowledge of particular taxa.

Moura and Jetz conducted an innovative study aimed at identifying taxonomic and geographic gaps in our knowledge of species, directing efforts to groups and places where the greatest returns on investment seem likely. Their data focused on vertebrates, but their approach could be generally applied. They identify amphibians and reptiles as two groups of particularly great opportunity and conclude that Brazil, Indonesia, Madagascar and Colombia, combined, might yield as many as 25% of the vertebrate species yet to be discovered. Uneven quantity and quality of data mean that some groups, including many invertebrates, are less amenable to their model, yet it points to one possible criterion for prioritizing field work.

Debates over priorities for each funding cycle will be intense, as they should be. Not surprisingly, everyone wants to go first. But success will come only if the community can agree to an explicit list of priorities and univocally support them. Vying for priority will bring forth the best ideas and arguments, and the quality of proposals will benefit from fierce competition. There should be an annual assessment of progress with minor adjustments to respond to advances, changing circumstances and emerging opportunities. The community should have a list of priorities ready to go at all times, so that no sooner is one goal reached than another takes its place in a rolling agenda.

Certain priorities can advance all taxa, to some degree, at the same time such as a cyberinfrastructure backbone, museum storage modernization and advanced instrumentation accessible to the community. Beyond taxa selected for special attention, the project should assure that all taxa have experts, somewhere, working on them at all times. Qualified, productive specialists, working on taxa within the purview of the inventory, should have access to competitive funds to support their research.

At the top administrative level, efforts should be made to establish partnerships and agreements that create a diversified base of funding. For example, in the U.S. taxa of importance to crops or medicine might find support from the agriculture department or national institutes of health, freeing up funds for basic work in other taxa, but this would likely require a mandate from on high (see below). With the promise of biomimetic breakthroughs, it is reasonable to expect investments from the private sector, interested in new ideas to advance industries in sustainable ways.

Rancorous theoretical fights in the 1970s earned systematists the reputation of being mean-spirited. I prefer to describe it as intensely passionate rather than vicious, but understand where the impression came from. The reduction of support for descriptive taxonomy has not been good for unity, thinking big, or planning long term. There is an old saying that academics argue so intensely because they have so little to fight over. This is sadly the state in taxonomy.

Like the astronomers, we need a governing body to oversee the inventory, prioritize projects, coordinate participants, manage funding sources and hype success stories along the way. When we proposed in *Systematics Agenda 2000* that various government agencies step up to fund taxonomy, each in their mission areas, rather than having an independent budget to drive species exploration, we were naïve. Agriculture did not volunteer to fund more than a few studies of groups of pests or potential biocontrol agents; NIH did not offer to aggressively fund the taxonomy of groups of biomedical importance. The core of an inventory must be structured as an enterprise on its own bottom, cooperating with agencies, institutions and organizations to the extent possible, but driving systematics, museum modernization and the inventory forward on its own resources and timetable. Agencies are invariably pulled in many directions and few have the luxury of making commitments for longer than a few years, unless Congress were to direct them to do so.

A worldwide inventory requires the participation of the world. A respected international organization representing systematics is needed to spearhead the effort and assure that it is not commandeered by other agendas. The United Nations could conceivably initiate global planning or, perhaps more promising, a group of national science academies. Or, as has been proposed in the past, a supersociety might be formed as a coalition of systematic biology professional societies that could coordinate community conceptualization of the project. Whatever the mechanism, organization will be a challenge. It will need to be tightly coordinated and managed to keep the aims of an inventory focused yet have the flexibility to accommodate diverse participants and rapidly changing circumstances.

As much as I am animated by science, let's face it: the success of an inventory ultimately depends on money. I would like to see the world's nations chip in to support the project through a combination of funds, in-kind contributions and debt forgiveness, but it remains that leadership for an initial phase of creating an inventory will likely require one or a few visionary organizations to start the process. Those capable and willing to commit core funding sufficient to initiate and sustain the project would necessarily include national-level governments. As the benefits of the inventory became clear, the ideal would be an international consortium of such government agencies. And, I would hope philanthropic individuals and foundations would see in this an opportunity to grant something of incredible value to humankind by stepping up to support aspects of the inventory.

It makes sense to look to organizations with experience putting together similarly large, complex initiatives. I am thinking of the National Academy of Sciences in the U.S., the Royal Society in the U.K., and the European Science Foundation, as three examples, with similarly capable organizations in many other nations. In the U.S., a National Academy of Sciences study can have enormous influence on budget allocations and tap leading scientists to refine the plan. Whatever the first steps in creating a clear vision and goals, the implementation must include genuine international buy-in. From sharing costs to opening borders for exploration, every nation, large and small, has a uniquely valuable role to play and benefits to gain. The best hope for success engages taxonomists, museums, universities and governments around the globe as full partners.

Astronomers and physicists have been particularly successful in funding "big science" projects, including expensive shared instruments. Astronomers have successfully implemented generations of land- and space-based telescopes. We do not need a telescope in orbit, but we do need a sophisticated cyberinfrastructure and an internationally distributed network of natural history museums that are both active centers of research and the world's archives of biodiversity knowledge. An inventory is as grand a science project as any, simply differing in details, needs and implementation.

The Planetary Biodiversity Inventory projects, initiated when I was at the National Science Foundation, demonstrated benefits of collaborative taxonomy. Teams of experts and institutions increased the speed of species discovery. With the ambitious goal of rapid expansion of knowledge of species in targeted taxa, these five-year grants produced thousands of species new to science. Teamwork is a necessary component for the inventory, but not sufficient in itself. It is imperative that the research platform and infrastructure for systematics research be modernized. Cyber-enabled taxonomy needs to mature as a fusion of traditional descriptive taxonomy, phylogenetic systematics and all that information science and digital instrumentation can bring to the game.

I have always marveled at how productive taxonomists can be with few or no assistants and wondered what they might achieve given even modest support staff. We do not expect world-class chemists to wash glassware or order supplies yet have no difficulty expecting the world authority on a group of plants or animals to collect, prepare and label specimens; make measurements, drawings and digital images; plot maps; conduct literature searches; maintain databases; draft descriptions; and edit manuscripts, among many other routine activities that could be delegated in whole or part. It is imperative for an inventory that taxon experts are permitted to do revisionary and monographic work full time. Too many experts are expected to engage in some kind of popular science, pursuing taxonomy on the side. There are too many species, too few experts and too little time to dilute expertise in this way.

I am as worried about intellectual freedom for individuals as I am excited by teamwork. The intellectual depth and demanding scholarship of taxonomy naturally lend themselves to scientists working alone. A monograph, like any great creative work, is a very personal thing to the scientist investing years in its making. It is important that we balance the efficiencies of teamwork with the necessity to protect and nurture individual thought, creativity and credit. Even an e-monograph evolving incrementally with additions and corrections by many contributors will require periodic comprehensive revision in a manner analogous to traditional printed monographs. Mechanisms for recognizing and giving credit to such herculean efforts must exist, even as knowledge is consolidated into a community-maintained e-monograph. Surely, clever programmers can conceive the right kinds of metadata to keep track of individual contributions.

Major advances in science often have their roots in the work of one person who dared question the *status quo*. Leon Croizat was thought to be a nut by most of his contemporaries. His most important books were so contrary to popular thought that they could not pass peer review and had to be published privately. An eccentric, he appeared to have gone off the deep end compiling massive amounts of data about the geographic distributions of hundreds of entirely unrelated species. But there was genius in his apparent madness, and we now credit the science of vicariance biogeography to Croizat. His "general tracks," distribution patterns repeated among diverse groups of species, pointed to major geologic events affecting whole biota.

We must also make allowance for personalities. There are people who simply work best as an individual researcher-scholar, including those I suspect had the "does not play well with others" box consistently checked on elementary school report cards. Provisions should be made to fund independent researchers, allowing for minority views and unconventional ideas and methods to be explored. In systematics, as in life, it takes all kinds.

There is nothing more dangerous to the advance of knowledge than committees of experts being given the power to dictate which species or classifications are to be accepted, silencing minority opinions. The temptation to use such authoritarian methods to impose stability on names must be avoided to protect science and freedom of thought. The existing codes of nomenclature in botany and zoology do an excellent job of finding the right balance. The further design of software for descriptive work represents an opportunity to safeguard competing views and avoid taxonomy by fiat. This is analogous to deliberations by the U.S. Supreme Court where cogent arguments are memorialized for both majority and dissenting opinions. Thus, e-monographs should include layers of information with easy access to consensus classifications for casual users and the ability to drill down to dissenting views for experts.

The more we know about species, the better we will become at predicting where to look for new ones. There are many examples of mining monographs for clues about where unknown species might be found. Gaps on maps and knowledge of ecological requirements is another way to pinpoint likely places to search. Backing this up with satellite images that tell us the state of habitats, planning expeditions is becoming an art unto itself.

While I have argued that we keep room for lone researcher-scholars, it is critical to create centers of collaborative taxonomic excellence. Such hotspots of taxonomic activity would have a critical mass of professional systematists and support staff, undergraduate students, graduate students and postdocs, a world-class collection, state of the art research infrastructure and responsibility for maintaining one or more taxon knowledge-bases (i.e., e-monographs). Every natural history museum should develop such centers of excellence, reflecting their strengths and priorities. Museum educators and public relations staff should be integral parts of centers, disseminating knowledge to visitors, through the Web, exhibitions and earned media coverage.

Beyond an explicit plan for the first decade, with milestone accomplishments along the way, three initial goals could ramp up the capacity of the community to accelerate species discovery: (1) modernization of research infrastructure, with special attention to collections accessibility and cyber-enabled taxonomy; (2) organization of international teams of taxonomists and institutions to aggressively target selected taxa; and (3) creation of mechanisms to provide support staff and modest grants to each of at least 2,000 active taxonomists worldwide.

The aim of this three-part investment scheme is to increase the number of new species named each year, thereby setting the inventory on course to reach 10 million species in 50 years or less. The 2,000 full-time taxonomists could potentially reach this mark on their own, but pursuing all three creates a safety net for meeting or exceeding the goal.

In the event of war, natural disaster or the loss of political will to continue funding collections, it is risky to put all specimens of any taxon in one geographic location. There is still a bombed-out shell of a museum building in Berlin and shrapnel marks on the walls of the Natural History Museum in London. Modest redundancy makes for good insurance, duplicating species holdings at multiple sites. That said, there ought to be centers where all, or nearly all, species are represented, used in research and curated by an active, world-class team of scientists, curators and technicians. This will require a new kind of coordination among museums. Institutions must dare to focus on more than a few charismatic groups, like mammals, birds and dinosaurs, so that every major lineage has one or more centers of excellence.

For species discovery, general collecting has serious limitations. It's more like a lottery than a strategy. When the goal is to know all species, there is no substitute for specialized knowledge and field experience. One of my students had become the world authority on an obscure family of slime-mold-feeding beetles. He traveled to a well-known site in Central America where general insect collecting techniques had been used for decades. In all that time, and in spite of assembling a respectable-sized collection, not a single specimen of the family he studied had been collected. In a few days, he collected five species, including some new to science. With specific knowledge of these beetles, and experience ferreting out micro-habitats where their presence was possible, he was able to zero in as no general collector could. Most such stories involve a taxon expert going into the field, and this practice must continue. But to speed up discovery, it makes sense to create specialized teams of collectors, too, who can similarly identify common species on sight, avoid redundant collecting and focus efforts on potentially new species.

Rebuilding the kind of taxonomic programs that once existed in universities will take time. In the meanwhile, and to address the need to engage citizen scientists in the effort, I propose a different approach to taxonomic education as an immediate solution. The best minds in systematics should be invited to participate in the creation of an Open Linnaean Academy, an open-access, online systematics curriculum. Ideally including live sessions with the opportunity for group discussion, they should in any event be recorded and made available at the convenience of students. Amateurs, and students who aspire to a career in taxonomy, could have access to world-class lectures on systematics theory and practice. The curriculum should include general courses to build a conceptual foundation for doing taxonomy well, and specialized classes focused on individual taxa at multiple levels. The latter includes practical instruction in making identifications. Initial steps in this direction exist, such as the Distributed European School of Taxonomy.

Because we need all the help we can get, this taxonomic academy should be open and free to anyone interested in learning. Making it free, we can encourage a higher level of competence among citizen scientists, amateur collectors and beginning students. Auto-tutorial curricula and self-administered online exams should also be freely available so that students may pace themselves and assess their learning. For those wishing to be credentialed in systematics, there ought to be a fee-based pathway that includes examination by an expert and the option of badges, accreditation or academic credits in partnership with accredited universities and museums. Additionally, funds should be made available for aspiring taxonomists to spend time with established authorities in the group of their interest. There exists knowledge and techniques not recorded anywhere that should be transferred to a new generation. With an aging taxonomic workforce, such apprenticeships should be a high priority.

Recently, Ziegler and Sagorny used digital imaging technologies to describe the internal organs of the only known specimen of a new species of deep-sea octopus without destructive dissection. Unlike the type specimen, their digital images are immediately available to researchers worldwide. While digital representations of type and rare specimens are immensely important, they can never eliminate the need to study actual, preserved museum specimens. Physical specimens contain more information than we know about or can fully document in any medium. New characters will continue to be discovered prompting reexamination of specimens. Many characters, to be precisely interpreted, must be viewed from particular angles or with special lighting. It is not uncommon that a specimen must be repositioned or dissected to make a detail visible.

Access to type specimens has always been an obstacle for taxonomic progress. For many purposes, high-resolution images may prove sufficient. But physical type specimens will always remain the last court of appeal for decision-making. Engineers should be engaged to design automated imaging systems that produce high-resolution images. A priority ought to be the creation of e-types, that is, composite, digital images of type specimens. Over time, experts can augment a growing collection of images so that they show all critically important characters. This would mean that taxonomists anywhere could quickly consult images of type specimens to assure that names are being used correctly.

Museums and botanical gardens have begun this process in a big way. Prime examples are the herbaria of the New York Botanical Garden and the Museum National d'Histoire Naturelle in Paris, which have systematically scanned all of their plant types. Depending on the characters that must be examined, a range of technologies can be applied. Most plant specimens are on flat herbarium sheets and features can be seen from high-resolution, 2D images. Pinned insects ought to be scanned as 3D images, as well, so that they can be rotated on multiple axes. And vertebrate specimens should be scanned with computer-assisted tomography, so that skeletal characters can be examined. It cannot be overemphasized that digital images, no matter how useful or seemingly complete given current knowledge, must never be understood as acceptable substitutes for actual specimens. They merely offer quick, efficient access to virtual specimens that, under certain circumstances, interpreted by an expert with appropriate knowledge, can affirm the presence or absence of an attribute, or confirm that a scientific name is being used properly.

It is disgraceful that it can take up to two years to learn about newly described species in zoology. Innovative approaches to registering new species and nomenclatural acts have been proposed, most conspicuously by IPNI (the International Plant Names Index) and ZooBank. What needs to happen next is to require registration before new species names or name changes are valid. This would assure that all nomenclatural acts, including newly named species, are available to the community immediately and can be located through one, open access web site. It should be obvious that registration must not include censorship of any kind, merely conformity to established rules and standards.

From time to time, renegade, unqualified amateurs, in groups as diverse as spiders and flowering plants, clog the pipeline by naming large numbers of poorly conceived "new species," presumably in a quest to see their own names in print. These are often no more than slight genetic variations, yet rules of nomenclature dictate that even frivolous names must be dealt with. There is a brief history of this "*mihi* itch" by Evenhuis that, in the words of Macleay, is "that morbid thirst for naming new species."

Scores of initiatives in recent decades have claimed to address the so-called taxonomic impediment, i.e., the inability to identify most species. Most have involved workarounds failing to address the needs of taxonomists. Users are ultimately best served when taxonomy is done to its highest standards. Therefore, the approach going forward should be to address the requirements of systematists to work efficiently, then assure that discoveries and conclusions are translated into a form accessible to non-specialists. Simply moving outdated knowledge onto web pages is barely progress. We need to support taxonomists to do taxonomy and to create and repeatedly test taxonomic knowledge.

I don't want to leave the impression that there are not many committed individuals and organizations around the world dedicated to a taxonomic renaissance and species inventory. Taxonomists are keenly aware of the crisis of support for fundamental taxonomic studies. For example, a decadal plan was developed by the Australian Academy of Science and Royal Society Te Aparangi (New Zealand) that envisioned a "hyper-taxonomy" approach to an inventory of the 70% of species not yet known in their region. But an undeserved image problem and bias toward experimental biology continue to hold systematics back. There is far too much at stake in the current crisis to permit this misguided character assassination of taxonomy to continue.

Recently, some have criticized the practice of collecting specimens for museums. These are generally individuals who work on vertebrates with small populations for which even modest collecting can be a threat to species survival, or molecular systematists who see no further use for morphology. Tight restrictions on collecting are clearly appropriate for threatened populations. But for millions of others, including vast numbers of plants and invertebrates, scientific collecting, with few exceptions, poses no existential threat and is necessary for detailed observations.

Rather than collecting, some propose to sample only DNA, either from the environment or from non-lethal extractions from living specimens. Others suggest a photographic catch-and-release approach. Neither can tell us very much of what we need to learn about species and collecting remains the only practical alternative for a massively large number of species. We know so few species that it is absurd to think we can say which characters will be sufficient for telling them apart when they are all known.

Further, without voucher specimens in museums, there is no way to repeat observations. DNA constitutes a very narrow subset of a species' characters. And even the best photography cannot capture all important features of a specimen. With a mass extinction underway, it is no longer safe to assume that specimens can be collected at a later date. For millions of species, museum specimens are the last best hope for studying them in detail, and long term.

The ethics of collecting are complicated, without doubt. But there are ethical issues attached to failure to collect specimens, too. It is incredibly arrogant to presume that we know as much as we might ever want or need to know about species and therefore cease collecting them. In recent years, an unexpected second thymus gland was discovered in the laboratory mouse, one of the most intensively studied species. Only by collecting and preserving specimens can keep open the possibility of such new discoveries. They will not come from DNA data or photographs. A combination of science, sensitivity and common sense can guide us to the right balance between conservation and collecting.

With all good intentions, there exist laws and regulations that indiscriminately impede scientific collecting. No right-minded person would defend elephant poaching or orchid fanciers collecting rare species for their personal amusement. But scientific collecting is another matter. A team of legal scholars should be assembled to address issues of legitimate scientific collecting, with reciprocity between granting access to habitats and sharing resulting knowledge.

In the past, examples of economic activity resulting from species exploration were rare beyond a few highly publicized pharmaceutical discoveries. It is now foreseeable that biomimicry will change that. Any species might suddenly emerge as the inspiration for new products, designs or materials, possibly spawning entirely new industries, and we must get ahead of the inevitable question of who profits. A flood of new restrictions on collecting, aimed at guarding against financial exploitation, would only slow the inventory, stifle advances in biomimicry and impose greater ignorance on us all.

Ideally, knowledge of species ought to belong to humankind. It is naïve, of course, to believe unrestricted worldwide collecting will be welcomed when potential fortunes are at stake. So, integral to planning an inventory is addressing intellectual property rights. There are precedents from which to work, such as the OAS *American Declaration on the Rights of Indigenous Peoples*, the United Nations Convention on Biological Diversity that addresses bioprospecting, the World Intellectual Property Organization and a growing body of international law. There must be recognition that many species have no economic value, while others do or may in the future. Contingencies for equitable sharing of profits, statutes of limitation and related issues must be ironed out among participating countries. But allowing millions of species to go extinct undiscovered profits no one.

For now, a priority is to remove as many obstacles to species exploration as possible to make way for an inventory. This means licenses to collect for scientists or their agents, for scientific purposes, with the deposition of specimens in publicly accessible museums. Perhaps the governing body, representing the interests of participating countries, could issue, coordinate and assure compliance with collecting permits for the project. While tight restrictions on collecting protect wildlife, they also prohibit the growth of taxonomic knowledge and everyone suffers as a consequence. Countries that want to develop their species resources should be first among those supporting credible scientific species exploration. Species that go extinct unknown will make no one rich, and all of us poorer.

Species exploration builds on centuries of accumulated specimens and knowledge. Taxonomists and students in the most species-rich regions of the world are usually at a disadvantage because so many specimens collected in their countries are housed in museums in Europe and North America. Although politically incorrect and painful reminders of colonialism, for scientific reasons it is useful to keep these specimens where they are so that they can be compared side-by-side. Fortunately, digital technology offers a compromise. Scientists and students can be given priority access to type and rare specimens from their countries as a form of virtual repatriation. This would include both image archives and scheduled, direct access to specimens via telemicroscopy. Social and political circumstances have changed dramatically since the days when European powers gathered specimens with wild abandon from colonies around the globe. Addressing the issue of repatriation of existing species knowledge can help build trust that, as new species are discovered, they and their information content will be accessible to everyone, anytime, anywhere. Going forward, great effort should be made to duplicate specimens to the extent possible. A good model would be specimens in one or two synoptic world collections, functioning as centers for taxonomic research, paired with comprehensive collections of regional flora and fauna housed in national museums.

Paradoxically, although launched by Linnaeus in the 18th century, a global-scale project is something relatively new for biology. As progress is made in the inventory of species, there is no doubt that valuable lessons will be learned leading to new practices and technologies that further increase efficiency. The project must keep its eye on the ball, remaining focused on the goal of an all-species inventory. At the same time, society will be facing new and emerging challenges that will need to be addressed. Thus, the project should be open to adjustments along the way that both support its core mission and are attuned to environmental and other issues as they arise. The uncharted confluence of taxonomy, information science and biomimetic engineering will bring opportunities and challenges we cannot yet imagine. As knowledge grows, the impacts of the inventory will become clear to new and broader audiences, increasing its support.

An additional book would be necessary to flesh out a detailed plan for the inventory. A realistic blueprint, however, must take into account the aspirations and challenges of individual participants, whether individuals, institutions or countries. No one size will fit all taxa, or all participants. This is why scientists from around the world must convene to work out the specifics such that they respect individual priorities and needs and, in doing so, achieve buy-in. Below, I present elements of an inventory that seem particularly important. This is meant only as a conversation starter. It will be up to the systematics community, speaking through an international coordinating body, to work out the details and sequence of priorities.

Twelve Elements of a Taxonomic Inventory of All Species

1 **Clarity.** Clarify and make explicit the mission, goals and tenets of taxonomy, including its vision for a planetary-scale all-species inventory and classification. Embrace productive collaborations while guarding the independence of taxonomists, restoring a focus on species and their characters, and reengaging Hennig's theoretical revolution.
2 **International Coordination.** Create a governing body to oversee and coordinate inventory activities with the advice and consent of participating nations, institutions and taxonomic communities.
3 **Workforce.** Educate and inspire a new generation of professional taxonomists while training and engaging citizen scientists in meaningful, hands-on contributions. Enable efficiencies of teamwork while jealously protecting individual intellectual freedom. Provide free and open online courses of study in both the theory and practice of systematics and the identification and classification of particular taxa, as well as reestablish formal curricula and graduate programs in universities and museums.
4 **Modernize research infrastructure.** Design research infrastructure to specifically meet the needs of taxonomists to do descriptive taxonomy faster and to the highest standards. This should include centers of excellence focused on particular taxa, networked advanced instrumentation and a cybertaxonomy research platform.
5 **Remove barriers to collecting.** Engage international legal scholars to minimize obstructions to collecting for the purpose of a world inventory; develop and make known best practices that protect endangered and threatened species while simultaneously assuring that museum collections are comprehensive; assure that all specimens and associated information are openly and

freely available to all; confront the challenge of intellectual property rights as we enter the age of biomimetic solutions.

6 **Disseminate knowledge and increase public understanding.** Restore taxon-focused curricula at the college level; promote public education regarding taxonomy, species and phylogeny through museum exhibits and programming; develop a strong "shameless self-promotion" public relations unit within the inventory to celebrate discoveries and their impacts.

7 **Re-envision access and use of collections.** Modernize collection storage to balance physical protection of specimens, ease of access and maximization of information content. Create a comprehensive archive of "e-types" and centers where e-types may be created. Continue trend of opening museum collections and research to the public.

8 **Modernize codes of nomenclature.** Descriptions of new species and nomenclatural acts should be immediately and freely available to anyone. Revise codes to require registration of all nomenclatural acts. Provide scientific name verification service to journals who, in return, track use of scientific names, crediting original authors. Make e-monographs one-stop-shopping for what is known about taxa, including competing ideas, and real-time user-designed "publications" from identification guides to maps, checklists, etc.

9 **Establish centers for nature-inspired design and engineering.** Whether in universities, museums or institutional consortia, establish centers where taxonomic knowledge is mined to identify biomimetic models. Centers should serve as think tanks and prototype studios engaging taxonomists, information scientists, engineers, entrepreneurs, representatives from industry and others as appropriate to find sustainable solutions.

10 **Create funding model.** With the goal of a minimum expenditure of U.S.$1 billion per year for a period of 50 years, build a collaborative model that gathers and coordinates monetary resources for the inventory. Pull together a combination of core and peripheral funding, including public and private sources and in-kind contributions.

11 **Understand and strengthen human–nature connections.** Partner with sociologists, architects, city planners, historians, philosophers, psychologists, economists and others to improve our understanding of the importance of human connections to nature and build public appreciation for what can be saved, and gained, through an inventory and biodiversity conservation.

12 **Post-inventory taxonomy.** Because taxonomic knowledge is built on hypotheses, it is critical that taxonomic research not end with an initial, intensive species inventory. A continuing, vibrant taxonomic enterprise, at a lower level of intensity, will be necessary for conservation, biomimicry and understanding of our planet and its history.

A revitalized systematics and world inventory will lead to a scientific achievement so immense that few would have dared dream it possible: a detailed snapshot of the millions of life forms of a planet as it transitions from one geologic epoch to the next. The implications of such knowledge are unprecedented in the annals of science and will be transformative.

Further Reading

Anon (1994) *Systematics Agenda 2000: Charting the Biosphere. Technical Report*, Association of Systematics Collections, Washington, 35 pp.

Australian Academy of Science and Royal Society Te Aparangi (2018) *Discovering Biodiversity: A Decadal Plan for Taxonomy and Biosystematics in Australia and New Zealand 2018-2027*, Australian Academy of Science, Canberra, 63 pp.

Bebber, D. P., Carine, M. A., Wood, J. R. I., and Scotland, R. W. (2010) Herbaria are a major frontier for species discovery. *Proceedings of the National Academy of Sciences*, 107: 22169–22171. doi:10.1073/pnas.1011841108

Blackwell, M. (2011) The fungi: 1, 2, 3... 5.1 million species? *American Journal of Botany*, 98: 426–438.

Craw, R. C. (1984) Leon Croizat's biogeographic work: A personal appreciation. *Tuatara*, 27: 8–13.

Dikow, T., Meier, R., Vaidya, G., and Londt, J. G. H. (2009) Biodiversity research based on taxonomic revisions — A tale of unrealized opportunities. In *Diptera Diversity: Status, Challenges, and Tools* (eds. T. Pape, D. Bickel and R. Meier), Brill, Leiden, pp. 323–345.

Erwin, T. L. (2004) The biodiversity question: How many species of terrestrial arthropods are there? In *Forest Canopies* (eds. M. Lowman and B. Brinker), Academic Press, London, pp. 259–269.

Hull, D. L. (2008) Leon Croizat: A radical biogeographer. In *Rebels, Mavericks, and Heretics in Biology* (eds. O. Harman and M. R. Dietrich), Yale University Press, New Haven, pp. 194–212.

Janzen, D. H. and Hallwachs, W. (1994) All taxa biodiversity inventory (ATBI) of terrestrial systems: A generic protocol for preparing wildland biodiversity for non-damaging use. In *Report of a National Science Foundation Workshop, 16-18 April, 1993, Philadelphia, PA*, National Science Foundation, Washington, 132 pp.

Luc, M., Doucet, M. E., Fortuner, R., et al. (2010) Usefulness of morphological data for the study of nematode biodiversity. *Nematology*, 12: 495–504.

Miller, S. E., Hausmann, A., Hallwachs, W., and Janzen, D. H. (2016) Advancing taxonomy and bioinventories with DNA barcodes. *Philosophical Transactions of the Royal Society*, B, 371: 20150339. doi:10.1098/rstb.2015.0339

Moura, M. R. and Jetz, W. (2021) Shortfalls and opportunities in terrestrial vertebrate species discovery. *Nature Ecology & Evolution*, 5: 631–539.

Pearson, H. (2006) Surprise organ discovered in mice. *Nature News*, 2. doi:10.1038/news060227-9

Pyle, R. E. and Michel, E. (2010) ZooBank: Reviewing the first year and preparing for the next 250. In *Systema Naturae 250: The Linnaean Ark* (ed. A. Polaszek), CRC Press, Boca Raton, pp. 173–184.

Rafael, J. A., Aguiar, A., and Amorim, D. de S. (2009) Knowledge of insect diversity in Brazil: Challenges and advances. *Neotropical Entomology*, 38: 565–570.

Raven, P. H. and Wilson, E. O. (1992) A fifty-year plan for biodiversity surveys. *Science*, 258: 1099–1100.

Wheeler, Q. (1995) Systematics and biodiversity: Policies at higher levels. *BioScience*, 45: s21–s28.

Wheeler, Q. ed., (2008) *The New Taxonomy*, CRC Press, Boca Raton, 237 pp.

Wheeler, Q., Knapp, S., Stevenson, D. W., et al. (2012) Mapping the biosphere: Exploring species to understand the origin, organization and sustainability of biodiversity. *Systematics and Biodiversity*, 10: 1–20.

Zebich-Knos, M. (1997) Preserving biodiversity in Costa Rica: The case of the Merck-INBio agreement. *Journal of Environmental Development*, 6: 180–186.

Ziegler, A. and C. Sagorny (2021) Holistic description of a new deepsea megafauna (Cephalopoda: Cirrata) using a minimally invasive approach. *BMC Biology*, 19: 81. doi:10.1186/ s12915-021-01000-99781032484396_

20 Hall of the Holocene

Richard Fortey said, "I would go so far as to say that you can judge a society by the quality of its museums." As we slide deeper into a mass extinction, how we treat our natural history museums will have much to say about our society. In an earlier chapter, we had a glimpse of what taxonomy might look like reimagined for this age of accelerated extinction. Now, let's envision a natural history museum for tomorrow. Picture a future in which museums have rediscovered their purpose, differentiating themselves by focusing on taxonomic research and collections. With a clarified mission, staff, programs and resources are all focused on documenting and understanding species diversity and evolutionary history. Intellectual critical mass has been created by hiring clusters of taxonomists as researcher-curators. Museums have become universally recognized as world leaders in species exploration and trusted sources of species knowledge. The growth, development and use of collections is the heart of the institution. Exhibits and lecture halls enlighten the public, as researchers create and share knowledge and mentor the next generation of taxon experts.

With their identity restored, museums are no longer hostage to fashions in biology. Instead, they attract visitors, recognition and funding by being at the leading edge of species exploration. No longer locked away in back rooms, collections are visible, dynamic research resources, valued as priceless treasures in a rapidly changing world. Transformed from followers to pioneers, museums create their own luck by demonstrating impacts, benefits and opportunities derived from collections to a population hungering for leadership.

Creating, curating and caring for the greatest assemblage of specimens and knowledge about species, museums have become the public face for biodiversity, evolutionary history and biomimetic breakthroughs. They are ground zero for species exploration and confronting mass extinction—*the* defining scientific and environmental challenge of the millennium. Collections, and the knowledge they contain, are appreciated as humankind's biological heritage. These centers of systematics excellence have become synonymous with knowledge of biodiversity, preserving evidence of earth's past and creating the best chances for its future. Discovering weird and wonderful properties of species, they are also the tinder box in which a biomimicry revolution has been ignited.

DOI: 10.4324/9781003389071-23

Figure 20.1 The Natural History Museum, London. Housing millions of specimens, such institutions are a primary research resource for systematics. Photo: the author.

Shaking off the image of a dusty attic, museums open their collections and research to the world, disseminating knowledge and allowing the public to witness taxonomic discoveries in real time. Far from stale collections, museums are seen as dynamic institutions exploring life on the only species-rich planet in the known Universe, a source of ideas for living harmoniously with nature and tangible evidence of the diversity and history of living things.

Before continuing our tour of this imagined museum of tomorrow, let's take a reality check of the *status quo*. We have reached a tipping point in the relationship between human civilization and the natural world. No institution can provide better guidance or leadership for finding a route forward than a reimagined natural history museum. With the need to live sustainably, only clues from other species promise timely success. Wilson's Half-Earth Project has established a hopeful destination for conservation, but only knowledge of species can guide us there.

History will judge us by what we do in response to the greatest extinction event in human history. In particular, whether we have the wisdom and foresight to seize the chance to explore and inventory the diversity of life on earth, before it is decimated. By inventory, I don't mean a simple list of names or database of DNA sequences. But a *taxonomic* inventory including detailed descriptions of morphology and as many other attributes of species as possible, organized in a phylogenetic classification, backed up by museum specimens, recordings, tissues, observations and other evidence as may be appropriate and possible.

Most natural history museum collections are in a period of dormancy compared to the growth called for by the biodiversity crisis. The impressive Darwin Centre II at London's Natural History Museum is an architectural masterpiece and a model of ideal conditions for the storage of specimens. But it is not large enough to hold the entire existing insect collection, much less accommodate the kind of expansion we should expect from a world-class museum. And its design physically separates taxonomists from collections, based on the cynical assumption that scientists in the future will spend their time in molecular labs and collections will increasingly become historical curiosities. This is exactly the opposite of what science needs and society deserves. Collection growth should be extremely aggressive, focused and purposeful. We should expand collections like there is no tomorrow, because for millions of species, there isn't. And design them so that there is little or no distance between specimens and systematists. Rapid expansion of collections must begin now, carry on for as long as the extinction crisis exists and continue thereafter at a rate dictated by needs and interests.

Ecologists and population geneticists look at a collection, like the 30 million insect specimens in London, and see deficiencies. I see one of the greatest accumulations of evidence of species diversity on earth and a taxonomic research resource of boundless potential; physical evidence that allows observations to be repeated and hypotheses tested; an extraordinary assemblage of specimens from around the globe that can be compared, side by side, including those from times and places that no longer exist. For many species, there are no more than one or a few specimens, meaning that we can draw no conclusions about their abundance in a habitat or measure the frequency of genes in their populations. But that is not what museum collections were designed to do. Ecology and population genetics are best carried out in the field, where as much data as necessary may be gathered. Collections are made primarily by and for taxonomists—*to do taxonomy*. Ecologists make no apologies for sampling indicator species and ignoring the rest. Population geneticists have no regrets about focusing on one species and neglecting the others, nor should they. Yet taxonomists are criticized for not using quantitative sampling methods or assessing gene frequencies—but why would they? These are as irrelevant to the systematist as a synoptic collection is to questions in ecology and genetics.

Collections support systematics research focused at and above the species level. Yes, taxonomists must assess variation to the extent necessary to disambiguate species, but it is, above all else, the ability to compare species-level characters that makes collections so valuable to the taxonomist. For some purposes, it may be enough to know whether a species occurs on a continent or not. Not every question requires data at the square meter scale of the ecologist, or the intensive sampling regime of the population geneticist. The ideal collection for taxonomy includes representatives of every species and specimens from as many places as may be sufficient to show significant genetic and geographic variation.

Pulling a drawer at random in the insect collection in London, you rarely see large numbers of specimens of any one species. But what you do see is even more fantastic. Species diversity on full display. Representatives of more than 660,000 different species are housed in London. That represents approximately two-thirds of every kind of insect known to exist on earth. In systematics, there is no substitute for the ability to examine and compare so many species in one place. It is therefore tragic that so few taxonomists are employed by natural history museums to put collections to their best and highest use during this time of unprecedented species loss.

Not even the largest museums, such as those in London, Paris and Washington, contain examples of every species. It is a shared, international responsibility to assure that all species are represented somewhere. The task of a planetary-scale inventory is massive, and massively expensive, so it makes sense to avoid unnecessary duplication or pretending that museum collections can, or should, be primary research resources for experimental sciences. It would be possible to indiscriminately operate blacklight traps every night and fill gallon jugs with millions of insect specimens. But many would be common species for which the expense of preparation and storage is unjustifiable. It is collecting informed by taxonomic expertise that allows a museum to grow in the most valuable ways. A natural tension exists between two goals: having as many species as possible in one place, making comparisons easy, and not having all specimens in one place where war, fire, flood or numb-skulled administrators could lead to their destruction.

Duplicates deposited in multiple museums must be, to the extent possible, routine. In general, botanists have been better at the commonsense practice of scattering duplicates among collections. On the other hand, preparing and preserving thousands of identical specimens from the same place, when scores or hundreds would do, is a waste of resources. The overarching challenge to the museum community is to assure that collections, taken as a whole, represent species diversity in nature as completely as possible. And to assure that enough duplication exists in the network to guard against natural or manmade calamities and to allow for destructive studies, such as dissections, when they are desirable.

If there were not an extinction crisis, the current growth of collections might be good enough. The American Museum of Natural History in New York, for example, grows at a respectable 90,000 or so specimens per year, an exemplary number in comparison to many other institutions. Appropriately, they are also leaders in preserving specimens in ways that maximize the possibility of retrieving DNA now and in the future. This, and advances in techniques for extracting DNA from historic specimens, adds to the already massive information content of collections. Any practice that increases the amount and kinds of evidence preserved is to be commended.

From an architectural perspective, how much larger must we plan for museums to be? There is no easy or accurate way to answer this question, but we can make a rough estimate of the brick-and-mortar challenge by extrapolating from a simple ratio. Assuming that existing collections are about right to reflect the 2 million known species, we need only multiply by five to arrive at a baseline number for museums with 10 million species represented. The Natural History Museum in London currently has about 80,000,000 specimens for all taxa. We should be planning, therefore, to add about 4 billion specimens over the next 50 years in London alone! I would caution that this number is a *lower* estimate, given that many named species are underrepresented in collections today. On the other hand, the majority of species yet to be discovered, and those most underrepresented today, thankfully consist of small-sized organisms. Among mammals, new species will predominantly be small-sized, too, mostly rodents and bats, so no need for many cabinets sized for elephants and antelopes.

The cost of such expansion over the next five decades is immense, not to mention storage equipment, associated laboratory and preparation spaces and additional staff. This expenditure is exceeded only by the much higher cost of failing to grow collections. That cost will be measured in extinctions that might have been avoided, massive and numerous gaps in our knowledge of evolutionary history and missed opportunities to document clues for sustainable solutions to challenges facing humankind. When all is said and done a few centuries from now, a complete collection of species will be regarded as one of the most priceless possessions in the world. It will be the Louisiana Purchase of biodiversity with incalculable returns on investment. How do you put a price on a window into a lost, biologically diverse world through which we can continue to observe, explore and learn?

Alarmingly, trends in museums are headed in the wrong direction. Most are not growing at levels seen in the opening decades of the 20th century, much less at a pace appropriate to confront the biodiversity crisis. Some have divorced research and curatorial staff. The motive was presumably financial, detaching research positions from collections so that they could pursue more popular (read: easily funded) kinds of research. This lack of commitment to the growth and use of collections has contributed to the problem.

There are no simple solutions to the kind of growth now called for. The Smithsonian, located on the mall in Washington, and the American Museum of Natural History on Central Park West in Manhattan are prime examples. To reach the public, they could not be at better addresses. But for growth, there are few worse locations. Real estate is expensive in city centers, even if contiguous properties exist for development. One obvious answer is to divide and conquer. Keep public exhibits and some collections and research on site and move other collections and staff to remote locations. With cyber-connectivity, this may become less objectionable in the future. But something incredibly valuable is lost when collections are split and when staff no longer informally interact.

It is easy to criticize museum decisions, but they operate in an incredibly challenging environment. Most do a good job protecting the physical well-being of collections, but when they are sealed off, untouched and unused by systematists, they gradually lose scientific value and reason to exist. Just keeping museum doors open is no small feat given operational costs and competition for the public's attention. With the added challenge to grow collections, it is time for museums to do some soul-searching. They must clarify the role they wish to play in this extinction crisis and evaluate their priorities, obligations and opportunities as collections-based institutions in the current circumstances.

We have enough things to worry about without focusing on our legacy reputation, yet I would hate my generation to be remembered as so short-sighted, selfish and superficial that it failed to seize the last opportunity to explore the diversity and history of life, before it was too late. Or settled for molecular data when evidence as diverse as species themselves was there for the taking. We must ask ourselves some serious questions. Do we have the resolve to explore the biosphere of this planet to create a permanent record of its species? Will we bequeath collections and knowledge of biodiversity to those who follow us, or will their inheritance consist of a denuded biosphere riddled with environmental problems? We can make this *the* century of biodiversity exploration with more species discovered than in all time, before and after. A worldwide inventory of species can support conservation and reduce species loss, preserve evidence of evolutionary history and open the floodgates of creativity for reimagining how civilization—and the natural world—can be sustained.

There are few places on earth where evidence of humans is not seen. In the farthest reaches of the oceans, most remote islands and deepest forests, the impacts of humans are present, from deforestation to air, water and plastics pollution. We have changed the world so extensively that scientists have concluded we are entering a new epoch defined by our own deepening footprint on the planet: the Anthropocene—the Age of Humankind. If this is so, it means that another epoch is drawing to a close. The Holocene began only 12,000 years ago, at the end of the last ice age of the Pleistocene. Aware of this transition, for the first time in history it is possible to preserve evidence of biodiversity before an epoch has passed.

When I was keeper of entomology at the Natural History Museum in London, I sometimes detoured through the Hall of Dinosaurs on the way to my office. I know why I love natural history museums, but I'm curious about what other people get from their visits. Some visitors linger, carefully reading every fact on the placards, while others take in the dinosaur sights and shuffle along. I concluded that, either way, visitors, young or old were sharing one common experience. For however long they dwelled, or how many facts they gleaned, they were transported in their imaginations to another time. A day in the Jurassic, millions of years ago, when dinosaurs ruled the earth. Beyond the towering stature of these beasts, one is subconsciously reminded that size, power and dominance are no match for natural selection or environmental change. Death is an ever-present possibility; a question of when, not if. As Jim Morrison put it, none of us get out of here alive—and that goes for species, too.

Geologic epochs represent chapters in the chronicles of life and the Anthropocene will not be the last. The length of our reign will depend on whether we learn to live harmoniously with surviving species and the processes of the biosphere; whether we detect and adapt to inevitable environmental changes beyond our control. If natural history collections are expanded to include all species, they become a snapshot of biodiversity as we found it. As extinction proceeds, these same collections will serve as our collective memory of life as it was. This recollection is important as a benchmark for conservation and restoration ecology goals, and an enduring reminder of where we came from. The extent to which we honor the memory of other species says a lot about who we are. Collections are the best way to assure that we continue to remember, and to learn the story of evolution in greater detail. Knowing that story should be a priority for us because we are part and product of it and, now, author of its next chapter.

Picture a group of school children visiting a natural history museum a couple of hundred years from now. Just as today's visitors are transfixed by dreams about the age of dinosaurs, they will enter the Hall of the Holocene and their imaginations will carry them to an equally wondrous age, not all that long ago, when life was incomprehensibly diverse. Living in a world whose biodiversity has been severely diminished, they will stand in awe before life in the Holocene. With millions and millions of extinct species on display, they will simultaneously grieve for what has been lost, be inspired by what was, and, perhaps, leave with a heightened appreciation for what remains.

I love historic architecture because its endurance is a stark contrast to the ephemeral nature of human life. You are less full of yourself recognizing that you are but the latest participant in a long parade of humanity. It is humbling to walk passageways in Oxford University, or stroll the Rue Mouffetard in Paris, realizing that you tread on stones walked upon by people a thousand years ago, or more. I am privileged to live in a pre-Revolutionary house built in 1728. For me, it is a constant reminder that the modern world that I take for granted has not been here very long, and neither will I. It gives me a subconscious nudge to treasure the things I love, appreciate those who came before me and make good use of the few years I'm allotted.

Museums play a similar role for biodiversity, not allowing us to forget the amazing life forms of the past or the fragility of species today. Such collections are, in any case, the heart of systematic biology and our conservation conscience. They hold the raw material for taxonomic and phylogenetic studies and stand as a permanent record of organismal diversity and history. They make what we think we know from observations of species repeatable. To the extent that we succeed in saving species, collections allow us to recognize, enjoy and study them. Alternatively, however many species are lost to extinction, a Hall of the Holocene will guarantee that we continue to admire and learn from species no longer found in nature. A comprehensive collection of species is not a biodiversity mausoleum. To the contrary, it is a dynamic source of knowledge, information and inspiration to sustain the living. As collections are assembled, the public should be invited behind the scenes to share in the excitement of discovery as it happens.

The entire collection representing the Holocene will be housed in many institutions, of course. It will consist of an international network of physical museums, as it does today. The difference being that nodes in this distributed, virtual mega-museum will be more closely connected and coordinated than ever before. Few people are aware of what goes on in the offices, laboratories and collection ranges of leading natural history museums, outside of the public eye. To the uninitiated, the museum can appear to be an aging assemblage of stuffed skins, pickled fishes and pressed plants, only a few of which are on exhibit at any time.

Before describing my vision for the natural history museum of tomorrow, I want to pay homage to those of the past.

Museum modernization is urgently needed, but it would be criminal to sweep away surviving charms of museums past. I personally love the romantic intrigue of Victorian curiosity cabinets. There is, perhaps, no better surviving example than the natural history museum in Dublin, Ireland. The locals affectionately call it the "dead zoo," which is at once amusing and apt. Artful taxidermy displays allow you to get up close to animals in a way impossible in a living zoo. I sometimes hear criticism that such displays have remained unchanged for generations and ought to be replaced by new ones with the latest whistles and bells, but this ignores the fact that such exhibits are new in the eyes of a fresh crop of children every year. For others, they recall warm feelings from their own childhood, when they first discovered these wonderful displays.

The museum in Dublin was opened with a speech by none other than David Livingston, of "Dr. Livingston, I presume" fame. It seems to have remained unchanged since. It is a museum of a Victorian museum, complete with mahogany-framed display cases, creaky wooden floors and heads seemingly mounted on every square inch of the walls. It would be a travesty to modernize rooms like these. We should honor such historic spaces while at the same time expanding and modernizing other collection and research areas of museums for the exciting work that lies ahead. While grasping the opportunity to rethink how natural history museums are best organized, accessed and put to use, we should not lose sight that they are cultural treasures, as well as scientific assets.

Natural history museums were traditionally centers of excellence for taxonomy. How could it be otherwise, given the dependence of taxonomy on collections, and *vice versa*? As museums abdicated their leadership in systematics, both science and collections suffered. The growth, development and level of use of collections are reliable indicators of the health of taxonomy, and things are not looking so good these days.

I believe the separation of curation and research has been a mistake, imposed by institutional leaders neither trained in nor appreciative of the unique connections between research, scholarship and curation in systematic biology. It has led to hiring researchers in museums who have minimal or no use for collections and who contribute similarly little to their improvement.

Protecting the physical integrity of specimens is of the highest importance, of course. But this is sometimes carried to such extremes that scientists cannot study specimens in detail which violates the reason they were collected in the first place. There must be a rational balance between conservation of specimens and advance of science. Museums have a responsibility to the scientific community to both safeguard specimens and maximize information connected to and derived from them. When collections are not curated to reflect the growth of knowledge, their information content decays. Species names become out of date, and their arrangement is inconsistent with phylogenetic classifications. In the museum, there should be no daylight between taxonomic research and museum curation. Now for another brief, fictional foray into the future.

A Day at the Museum

The year is 2058. The city, London. Traffic on Cromwell Road is bustling, as eerily silent electric automobiles whoosh by. As a long queue awaits admission to the museum, anticipation is palpable. What used to be seen as the site of relics on display is now recognized as mission control for a space-age exploration of life on planet earth. A new respect for biodiversity has emerged by embracing the goal of an inventory of species, the elements of the biosphere. The rapid pace of species discovery has transformed the image of the museum, giving its collections and exhibitions the excitement of immediacy. It is both archive of legacy knowledge and the place where major discoveries are made at a blinding pace.

The grand entry hall is a tribute to life on earth, with a panoramic overview of the diversity of species from microbes to mosses, moths and mammals. It is both a roadmap depicting the sequence of major branches of the tree of life, and a table of contents for the museum as a whole. A color-coded legend on a large floorplan directs visitors to the various exhibition halls: dinosaurs, insects, plants, fungi, and so on. Two, in particular, catch your eye, the Hall of the Holocene and Project Linnaeus: Mission Control. You plan the shortest route to visit each in order.

Stepping into the Hall of the Holocene you are stopped in your tracks. Your mouth hangs open as your eyes struggle to take it all in. The floor and walls in this cavernous space are papered with a bewildering number of specimens, representing as many species. Some are large and imposing, like a polar bear and palm tree. Others, small to tiny, include centipedes, cephalopods, mosses, flat worms, wasps and diatoms. At a distance, most are mere dots peppered across the walls from floor to ceiling like pixels in a newspaper photo. As you stand before the massive array of arthropods, you notice a touch screen that controls a digital microscope that moves precisely on *x*–*y* axes up, down and across the wall, stopping on command to magnify the specimen in its line of sight.

You move this eye-on-biodiversity over a millipede that lives in Africa. Zooming in, you see that there are two pairs of walking legs on each of its seemingly innumerable body segments. Like a busy railroad switchyard, dozens of these cameras are in motion as visitors train them on this, and then that, specimen. With the touch of a button, you can send a photo of what you are seeing to your cell phone, or access an electronic file with as much natural history information as you can digest: when the species was discovered, its status living or extinct, and its distinguishing features, habitat and relationships.

But wait! Like a late-night infomercial, there's more. All the specimens have already been meticulously digitally imaged in 3D. In addition to the roving cameras, you can slip under a visualization helmet, select a specimen, and manipulate a 3D projection that you can spin around to view from any angle, at any magnification. In front of you is that millipede, projected to appear 6 feet long!

It's all too much to take in. You think about which species you will focus on the next time you visit, and go to check out Project Linnaeus. Crossing its threshold, you enter another massive room, this one filled with video displays and digital microscopes all in motion as they are operated by scientists around the globe.

Up-to-the-minute statistics snake across long, narrow LED displays like restless Wall Street tickertapes. You see figures steadily increasing in real time, as the number of species described, or redescribed, in Project Linnaeus increases with the predictability of a ticking clock. There are, on average, more than 500 new species named every day of the year, in addition to species being redescribed. There are running totals for a bunch of numbers: how many species have been discovered to date in this museum, for example, and by the worldwide inventory since it began in 2025. Some exhibits feature the latest biomimetic breakthroughs, and the species that inspired them. Others put the actual specimens of newly discovered species on display, allowing you to be among the first humans to ever knowingly see these newly named life forms.

In the center of the floor are rows of tall glass cylinders, each about 3 feet in diameter and 8 feet tall. They contain interactive microscopes. Actual specimens of recently discovered species are mounted under their lenses, and you are invited to grasp a joystick and examine them for yourself. Others have a bright red "in use" sign illuminated. These are being manipulated from remote locations by taxonomists and, thanks to a high-resolution display, you can see what they are seeing, as they see it, along with an explanation of their research.

You notice an alcove labeled "Species Exploration as It Happens." On one side are live video feeds from a team of collectors deep in the Amazon. On the other wall are a barrage of high-resolution displays. A Brady Bunch of live video streams can be watched here or looked in on from your home computer. You listen in as taxonomists in São Paulo, New York, Berlin and Paris debate the significance of a recently collected specimen. You see the specimen in real time along with them. Most exciting, you are there at the Eureka instant when they conclude that a species new to science has been discovered.

You print out or send yourself photographs of specimens, or purchase models of them. The 3D scanned data make it possible to produce exact resin copies of museum specimens, blown up or shrunk down to any scale. You might choose a *Triceratops* skull sized so that you can hold it in the palm of your hand, or enlarge a beetle mandible to 6 inches in length. Such models allow you to take a piece of the museum home and are invaluable visual aids for teachers.

Commentary on Natural History Museum Leadership

An inventory of life on earth will rank as one of the most ambitious and impactful science projects of all time, never to be repeated. Such a momentous achievement will require great leadership. Much of this will come from the taxonomic community, of course, as well as a governing body overseeing the project. Universities will be charged with preparing the next generation of professional systematists, hiring systematists to rebuild curricula and graduate programs. But no single category of institution has so much responsibility for the success of an inventory as natural history museums and botanical gardens with herbaria.

Historically, natural history museums were ground zero for species exploration when it was happening organically, not as a coordinated, international project.

Too many museum directors today see the collections in their care as historical relics, burdens rather than opportunities. Reasons for this are numerous. For the most part, directors are recruited to be fund raisers and may have no background in systematics, or even science. This is not necessarily a handicap. I have known several non-scientist directors who were simply first rate, not least of all because they were bright, quick studies and listened attentively to advice from their scientists. But there is the potential for things to go horribly wrong when leaders do not understand the theories, traditions and *ethos* of systematics.

Trends in science funding are frequently driven by the latest technologies, topics that have emotional appeal like conservation and those that address immediate problems like an emerging disease. As museum research programs chase external money, they can drift far away from any meaningful use of collections. This schism between research fads and collections can further diminish the apparent value of collections in the eyes of an administrator who is evaluated by a board based on the amount of money brought into the museum. This can be tragic for biodiversity exploration and systematics. When museums mimic ecology and genetics institutes, important questions based on taxonomy go unaddressed.

Museums have small numbers of scientists compared to research universities. Major museums may have no more than a few dozen PhD researchers on staff, while a top tier research university has hundreds of PhD biologists. With so many faculty lines, universities can afford to take risks hiring faculty in emerging areas of research. Some trendy science pans out to be a long-term winner, others prove to be a flash in the pan with the researcher becoming, within a few years, what is known in academic circles as dead wood. With so few positions in museums, taking risks on the latest fad is irresponsible management.

Because museums own the collections upon which systematic biology depends, they have a special obligation to physically care for specimens and to assure that they are used in research to their highest purpose. Far from altruism, supporting active taxonomic research adds to and expands the information contained in collections, enhancing their value. Scientists who live and work in the middle of a world-class collection have a privileged position and, thus, a responsibility to conduct systematics research not so easily done elsewhere. It is ridiculous to hire an ecologist or geneticist in a museum if they can do their research as well in a university without direct access to a collection. This is not to say that their research is any less valuable. But it is to emphasize that the number of research lines embedded in world-class collections is extremely small. With so many species unknown or untested, we owe it to the world to make the most of those precious few positions.

In recent decades, museums have increasingly abdicated their role as leaders in taxonomy. Rather than doing the heavy lifting required of leaders, they have opted for the path of least resistance, conforming to pop science in order to more easily attract outside dollars. When museums follow in the worn path of research universities, they are relegated to being also-rans. How can a handful of researchers compete with entire departments in research universities? They can't.

There are individual taxonomists doing first-rate research in near isolation, but there is much to commend a critical mass of systematists asking similar questions about the world. Ideas must be challenged in order to expose and correct their weaknesses. With collections as the nucleus, museums are natural places for the advance of systematics theory as well as knowledge about species and higher taxa. A critical mass of systematists, each focused on their own taxonomic interests, creates an electrified, creative environment of vigorous debate accompanied by waves of evidence and counter-evidence. Asking the same questions about homology, species, phylogeny and classification serves as an enormous common denominator among taxonomists with diverse taxon knowledge.

During my early career at Cornell University, I had the privilege of heading the Liberty Hyde Bailey Hortorium for a few years as the search for a new permanent director was organized. The Hortorium is one of a kind, created by Bailey himself. It focuses on the taxonomy of cultivated plants and, among its many great achievements was the production of *Hortus* which became the "bible" for countless plant nurserymen and horticulturists. My appointment raised eyebrows in the botany community, as I was an entomologist with negligible knowledge of plants. I was, however, a systematic biologist, and so too were each and every member of the Hortorium faculty. We had a great number of common interests in taxonomic theory and practice, and I look back on those years as some of the most intellectually rewarding and creative of my research career.

I once received a letter from Richard Hoffmann, a world authority on millipedes. He shared with me his fond memories of days, early in his career, when the Smithsonian's natural history museum was one of the world's crossroads of taxonomic research. Time spent there brought you into contact with other taxonomists wrestling with similar theoretical and practical challenges. It was an intellectually charged cauldron of systematic thought. That is, as Richard put it, "before the Smithsonian became a bush-league university." This was his way of articulating frustration over museums that choose an eclectic research program rather than one laser focused on systematics and putting collections to their highest use. Natural history museums have a choice to make. They can be unquestioned world leaders in taxonomy, species exploration, phylogenetic studies and classification. Or, they can be second-tier research institutes making contributions to random biological fields for which major universities have preeminence. Having been a university dean and president, I have deep appreciation for the pressures on administrators to attract external monies. And there is no question that knuckling under to current fads is an easier, and more certain, path to success by this perverted metric. But should making money really be the primary goal of a natural history museum or a university? If that is your goal, why not a career on Wall Street instead?

There are fewer leaders than followers for a very good reason. Leading requires vision, courage, untiring commitment, a willingness to swim upstream and a stomach for taking risks as a trail blazer. Every fad these institutions follow today was started by someone, somewhere, who was an early advocate for an idea or technology before it became popular. Given the unique resources of museums—the collections—they begin as taxonomic leaders from a position of unparalleled strength, which makes their cowardice and hesitancy to act like leaders even less understandable. Investing in expertise to capitalize on, and enhance through research and curation, the information content of collections was a winning strategy until posing as an experimentalist and speaking in terms of modern genetics became *de rigueur* in biology, starting about the second quarter of the 20th century. The simple fact is that we need a range of life sciences as diverse as life itself, not blind conformity to mob rule.

Beyond failing to use taxonomic research to steadily expand, improve and learn from the information contained in collections, museums are missing the boat with public education programming and exhibits. Here, too, there is a dearth of vision and leadership. Parroting science already popularized by someone else, somewhere else, may get people in the door, but it is a betrayal of the unique role of museums in science and society. Their exhibits and programs are an opportunity to reach millions of people with knowledge they should have but do not know to seek. Museums should capitalize on the specimens in their collections, dangling them like bait for people who want to be in the presence of an actual dinosaur or a recently discovered species. Museums can and should be cheerleaders for a taxonomic renaissance and a global species inventory. They can build public understanding and support like no other kind of institution. Good leaders recognize and seize opportunities—great leaders create them. It is time for natural history museums and their directors to become great leaders.

Behind the Scenes

The ways in which museum specimens are stored and accessed have essentially remained unchanged for more than a hundred years. As we undertake the greatest period of expansion of museums in history, it is time to take a hard look at how museums can be designed, constructed and used to more efficiently meet the research needs they are now called upon to address.

Robotics and digital imaging have redefined the possible. Obstacles to rapid taxonomic progress can quickly fall away if we invest in modernizing museum collections and research infrastructure. Image archives of type and rare specimens, in combination with remotely accessible digital microscopes, will level the playing field for scientists around the world. There will always be a home court advantage for those working in a major collection, but disadvantages to others can be dramatically reduced. Teachers can illustrate lectures with the rarest of specimens. Any qualified expert, anywhere, anytime can have the option of verifying identifications by comparison to images of authoritatively identified specimens. While there are many species for which such visual verification does not work, in the hands of an experienced taxon expert such resources can suffice much of the time.

Museums can be equipped with deep-freeze rooms to kill pests; organized to reduce the distance between taxonomists and the specimens they study; and designed to tear down the wall between research collections and public education. It is an incredible opportunity to remake museums for the challenges they are now called upon to meet.

While physical designs for specimen storage and access facilitate taxonomic research, the fusion of taxonomy with cyberinfrastructure will transform both systematics and how museums are used by others. Comprehensive databases will allow ecologists to virtually reconstruct habitats by determining which species co-existed geographically, and temporally, before mass extinction. Biomimetic engineers and inventors will be able to predict where to find models for problem-solving. And conservation biologists will make fact-based decisions about the places to prioritize in order to save the greatest number and diversity of species.

Outreach, Education and Hands-On Citizen Science

There are two obvious areas for natural history museums to exploit that, to date, are only tentatively explored. First is to take full advantage of digital instrumentation and the internet to make exciting research being done in collections and at field sites accessible to anyone, both in real time and archived. Going hand in hand, the second opportunity is to educate the public about the excitement and relevance of systematics. Biodiversity represents the last, great unexplored realm on earth yielding fantastic discoveries with each voyage into the unknown. Details of the history of species, and the diversity of complex and remarkable morphology, remain among the most fascinating untold stories. And the potential of biomimicry to learn from species ways in which we can improve our own lives, without trashing the environment, is inspiring at a time when good news is desperately needed.

I cannot imagine a richer set of educational resources or opportunities than those available to museum staff developing public exhibitions and programs, yet these cannot be used to full advantage in the absence of taxonomy. A worldwide species inventory combines the romance of an age when sailing ships charted the four corners of the globe with the sense of boundless possibilities in space exploration. Sharing the progress and discoveries of a planetary-scale inventory, as it happens, will open more opportunities for museums than can be exploited. I have found general audiences to be fascinated by stories of the discovery of individual species. I can only imagine the kinds of engagement with the public's imagination possible after we have launched a NASA-like campaign.

Keeping It Real

Digital imaging of specimens, and cyber-taxonomy in general, can accelerate taxonomy or become a distraction. When digital images relieve taxonomists of routine sorting of specimens, help other biologists identify species or enrich morphological descriptions, they are all to the good. I have heard some non-taxonomists opine that, once all museum specimens are imaged, we can dispose of collections, saving huge amounts of space and money. This is a wrongheaded, extreme version of the idea expressed by others that we photograph specimens in the field and no longer collect them. Anyone engaged in systematics knows that it is impossible to capture all the information contained in a specimen by any single means, least of all by a photograph. Detailed work frequently requires repositioning body parts or making dissections to see anatomical structures unobstructed. Excellence in systematics will always depend upon well-educated, experienced professionals working with real specimens, constantly probing the limits of our knowledge. And biologists clamoring about the importance of repeatable experiments ought to be first to recognize that repeating observations is impossible unless specimens are preserved. The idea of cybertaxonomic tools is to relieve taxonomists of routine tasks, add efficiencies where it makes sense and minimize obstacles to accessing specimens, knowledge and colleagues. It is not to replace either taxonomists or collections.

Summary

A new generation of professional taxonomists is urgently needed to create the workforce and intellectual guidance for a worldwide species inventory, but institutional leadership is no less vital. Universities have important roles to play in research and education, but no class of institution is more central to a successful inventory than those maintaining collections. As keepers of the world's physical evidence of species, they have an obligation to systematics, too. Instead of operating also-ran conservation programs, museums can create and verify the fundamental knowledge and information required to make the conservation of biodiversity a measurable success. Rather than mimicking ecology institutes, museums can be the source of knowledge that empowers ecologists to drill down to the level of species-to-species interactions. Instead of regurgitating science accomplished elsewhere, museums can use in-house systematists to create news. The small number of scientists in museums means that they cannot be all things to all people and still achieve great things. Museums have a simple but profound decision to make: continue in a supporting role for a potpourri of scientific priorities, or reclaim their stature as world-class centers of excellence in systematics. The world needs an inventory, now. An inventory depends upon collections, and a vibrant taxonomic community developing and using them. It is time that institutional leaders of vision and courage step forward to realize the scientific, societal and intellectual potential in collections. Time that systematists and natural history museums become masters of their own fate by accomplishing great deeds and sharing them with the world.

Further Reading

Amorim, D. de S., Santos, C. M. D., Krell, F.-T., et al. (2016) Timeless standards for species delimitation. *Zootaxa*, 4137: 121–128.

Asma, S. T. (2003) *Stuffed Animals and Pickled Heads: The Culture and Evolution of Natural History Museums*, Oxford University Press, Oxford, 302 pp.

Bakker, F. T., Antonelli, A., Clarke, J. A., et al. (2020) The global museum: Natural history collections and the future of evolutionary science and public education. *PeerJ*, 8: e8225. doi:10.7717/peerj.8225

Bebber D. P., Carine, M. A., Wood J. R. I., et al. (2010) Herbaria are a major frontier for species discovery. *Proceedings of the National Academy of Sciences*, 107: 22169–22171.

Davey, C. (2019) *The American Museum of Natural History and How It Got That Way*, Fordham University Press, New York, 278 pp.

Dorfman, E., ed. (2018) *The Future of Natural History Museums*, Routledge, London, 247 pp.
Evenhuis, N. L. (2008) The "Mihi itch"—A brief history. *Zootaxa*, 1890: 59–68.
Fortey, R. (2008) *Dry Storeroom No. 1: The Secret Life of the Natural History Museum*, Alfred A. Knopf, New York, 335 pp.
Girouard, M. (1981) *Alfred Waterhouse and the Natural History Museum*, Natural History Museum, London, 64 pp.
Grande, L. (2017) *Curators: Behind the Scenes of Natural History Museums*, University of Chicago Press, Chicago, 432 pp.
Kemp, C. (2017) *The Lost Species: Great Expeditions in the Collections of Natural History Museums*, University of Chicago Press, Chicago, 256 pp.
Macleay, W. S. (1838) On the brachyurous decapod Crustacea. Brought from the Cape by Dr. Smith. In *Illustrations of the Zoology of South Africa*, Invertebratae, IV (ed. A. Smith), Smith, Elder & Co., London, pp. 53–71.
Parker, S. (2010) *Museum of Life*, Natural History Museum, London, 192 pp.
Sheets-Pyenson, S. (1988) *Cathedrals of Science: The Development of Colonial Natural History Museums During the Late Nineteenth Century*, McGill-Queen's University Press, Montreal, 144 pp.
Wheeler, Q. (2016) This struggle for survival: Systematic biology and institutional leadership. In *The Future of Phylogenetic Systematics: The Legacy of Willi Hennig* (eds. D. Williams, M. Schmitt, and Q. Wheeler), Cambridge University Press, Cambridge, pp. 469–478.

21 Shameless Self-Promotion

Misperceptions about taxonomy, bias in favor of molecular data and experimental projects, and institutional preferences for science fads create a strong headwind for a systematics comeback. Beyond educating systematists, planning an inventory and modernizing infrastructure we must change perceptions. With few champions, the task falls primarily on systematists themselves. Changing minds within the scientific community will not be easy. It has taken decades to gnaw away at taxonomy's reputation and, realistically, it will take years to fully restore it. Two parallel tracks should be pursued to begin to improve matters. First, a major investment in taxonomy will entice institutional buy-in, if only for financial reasons. And second, an appeal must be made to the general public who, unjaded and apprised of what is at stake, can become important allies.

In restoring the reputation of systematics, a primary tool will be communication. Systematists must explain to everyone how rapidly species are disappearing, the implications of a mass extinction, the intellectual and practical benefits of knowing species and the fantastic opportunities that exist for taxonomy. As species are discovered, the story of evolution told, the state of biodiversity assessed and biomimetic solutions forged, communication of successes can lead to more and greater ones.

Regretfully, systematists have been their own worst enemies. Going along with the crowd, being willing participants in the subjugation of taxonomy to other, supposedly more modern, sciences has failed to earn respect or secure funding. Quite the opposite. It has created confusion about what, exactly, systematics is and undermined its ability to compete for positions and funds in pursuit of its own goals. When taxonomy pretends to be genetics, or any field other than itself, it is rightly seen as second rate. Why would those funding research invest in a runner-up version of a science, when they can support the real thing? Such a decline in funding is what happened when taxonomists emulated population genetics in the 1940s and tropical ecology in the 1970s. Passing taxonomy off as molecular genetics—as opposed to integrating molecular data into systematics—will bring more of the same.

To the extent that taxonomists have entered the public relations arena, they have largely failed. Now, with backs to the wall and wolves at the door, it is

DOI: 10.4324/9781003389071-24

imperative that we learn to be unapologetic, effective, forceful promoters of our science. We must be crystal clear about our mission and its benefits for the simple reason that no one else can or will do it for us. A public communications plan is as important to the success of a species inventory as engaging with museums and inspiring participants to get involved. Tongue in cheek, I often speak of the importance of a Department of Shameless Self-Promotion. Too often, we assume that exciting and important discoveries will speak for themselves. To some extent they do, as evidenced by occasional stories in the news about species discoveries or biomimicry inventions. But the number of such stories, compared to the flood of discoveries each year, suggests that there remains much to be done to raise the profile of systematics.

Self-promotion is seen by many in the scientific community as lowbrow. It simply isn't done. Or, when done, it results in dry, just-the-facts news releases. Don't get me wrong, this is not a complaint about public relations experts. Just the opposite, it is an indictment of taxonomists failing to be good partners for them. I have worked with, and learned from, exceptionally talented PR people, and there is no substitute for their knowledge, media contacts and instincts. Yet it remains that species experts are the ones with deep knowledge of discoveries, and their engagement is pivotal. At a minimum, taxonomists need to cultivate the ability to sense which discoveries are news worthy then take the initiative to share them with their institution's media relations office. In a routine manner, systematists need to make themselves available for interviews where they can share a bit of the excitement of species exploration and put a human face on systematics.

Promoting taxonomy should be an enjoyable chore. In my experience, people are interested and eager to hear stories about newly discovered species. I found a new species of slime-mold feeding beetle in a little woodlot, Smith Woods, on the fringe of Trumansburg, the village I lived in near Ithaca, New York. Interest in the species, mostly because it was found so nearby, took me off guard. I gave talks in public schools and for the Rotary club. It made local news. And a likeness of the beetle, *Agathidium aristerium*, was printed on an "Ithaca dollar," an upstate bartering currency.

Over the years, I have given many public lectures on the biodiversity crisis and the role of taxonomy in confronting it. Each time, I have been impressed by the level of interest and intelligent questions from general audiences. People truly care when the situation is spelled out for them, and I believe ginning up public support for systematics, museums and an inventory will be much easier than many suspect.

Taxonomists have been downtrodden so long that it no longer seems to occur to them that they are sitting on a scientific and media goldmine. There are few species about which some interesting evolutionary or natural history story cannot be told. Making species known to the public awakens something deep within each of us, a subconscious sense of the historical tethers that bind us to all other species. Systematics reconnects us to long-lost relatives and empowers everyone to experience species for themselves.

Museums can offer the opportunity to be in the presence of newly discovered species. But instead of sharing them with the public, they are unceremoniously locked away in back rooms. Without institutional administrators with the vision and courage to lead, we are left with feckless followers. They simply repeat announcements of discoveries made elsewhere, lifted from existing headlines, many of which have little or nothing to do with collections or systematics. During a calm, uneventful time, it might be tolerable to permit museums to remain secondhand storytellers. But in the middle of the greatest extinction crisis in recorded history, it is imperative that these institutions answer the call to leadership and shout out discoveries being made under their own roofs.

Universities bear a good deal of blame for the state of systematics, too. Taxonomy used to be a required part of the curriculum for every biology major, and systematists valued members of every faculty. Large universities, especially land grants in the U.S., had strong taxonomic programs in botany, entomology and zoology departments. When I was a student at Ohio State, majoring in entomology, there was a smorgasbord of taxonomy classes. I took not only insect morphology and introductory taxonomy classes, but also advanced courses on Coleoptera (beetles), Diptera and Hymenoptera (flies and wasps), Lepidoptera (butterflies and moths), smaller orders of insects and historical biogeography. Virtually none of those classes exist today to inspire and prepare the next generation of species explorers.

For several years, I wrote a column for *The Guardian* newspaper in London. Each week, I would write about a recently discovered species that was remarkable in some way. The response was impressively favorable. And, for a decade, I organized an annual Top 10 New Species list that received national and international media attention. I had been compiling the numbers of new species named each year and making colorful pie charts, but reporters showed no interest in statistics. But when I presented discoveries as a Top 10 list, the press was all over it. I see no downside to spicing up our reporting, so long as we remain scientifically accurate. In interviews for newspapers, magazines, television and radio, I had the opportunity to explain the dimensions of the crisis we face and share some of the excitement of discovering species. The visibility of the Top 10 led to an invitation to write the London newspaper feature, and a major publishing house inviting me to write a book with my personal favorites among recent species discoveries.

Below are a few ideas for promoting awareness of the biodiversity crisis and importance of taxonomy. While these and other initiatives need to be woven into the inventory plan, there is no reason to wait. By engaging in communications we can prepare the way for an organized project.

Top 10 New Species

Recognizing a list of the "Top 10" most unusual, beautiful, cool, repulsive and record-setting species each year draws attention to taxonomy, the biodiversity crisis, and can become one public face of the inventory when it is underway. So far, the Top 10 has consisted primarily of a news release, web site and interviews. It could be further embellished with a high-profile ceremony, perhaps hosted by a celebrity, as well as traveling exhibits including both posters and actual specimens of recent finds. And, of course, in the context of a global species inventory, it could be used to highlight the most impressive discoveries each year. I am currently working to reboot the Top 10 list and am aware of similar lists issued each year by several organizations.

Species Hall of Fame

To emphasize the impact of species exploration and taxonomy on science and society, there should be a Species Hall of Fame. Each year, authors of inducted species, if living, could be invited to a gala and recognized, a traveling exhibit produced to move among natural history museums and botanical gardens, and an elaborate archival web site created as a virtual, permanent Hall. All species, present or past, would be eligible. For example, an inductee might be *Zea mays* based on the agricultural importance of corn, or a newly discovered species that has spawned a biomimetic innovation or revealed an unexpected evolutionary insight. Scientists with expertise regarding the impact of an inducted species would represent it when authors are unavailable. The idea is to draw attention to important things that have happened as a consequence of discovering species, from advances in science to solutions to problems faced by humankind. And to recognize the taxonomists, museums and science behind the species names.

A number of categories seem obvious: Agriculture and Forestry; Oceans and Fisheries; Ecology and Ecosystem Science; Biodiversity and Conservation; Genetics, Development and Evolutionary Processes; Phylogeny and Evolutionary History; Palaeontology and Palaeoclimatology; Biogeography and Historical Biogeography; Systematics Theory; Biomimicry; and a catch-all At Large category. In addition to species, inductees should include taxonomists and others who either discovered species or promulgated their impacts. Among deserving human inductees are both historical figures (e.g., Darwin, Wallace, Owen, Cuvier and Linnaeus), and more recent ones (e.g., Willi Hennig, Gareth Nelson, Norman Platnick, Peter Raven and E. O. Wilson).

Making the Hall retrospective, we can go back in time to acknowledge species that have shaped history, society and science, sharing the amazing, often surprising, stories of the impacts of knowing species. A Species Hall of Fame could have a tremendous impact because it speaks to such broad issues and reminds us not only of past consequences of species exploration but the enticing potential in continuing the search for species.

I have been producing a weekly podcast and newsletter titled the Species Hall of Fame, but the concept could be more impactful if implemented on a larger scale, perhaps including a physical home like various sports halls of fame, as well as a web site as a clearing house for the impacts of taxonomic knowledge.

Celebrity Names

Our culture worships celebrities from movies, music and sport. There are celebrities with television shows whose only claim to fame is being famous. We need to take the hint and capitalize on this craze to bring attention to species and systematics. When Kelly Miller and I published our monograph on *Agathidium* beetles in 2005, we were unprepared for the avalanche of interest generated by a few celebrity names we created. Among the 65 new species were beetles named for our wives and Darth Vader—for entirely different reasons I assure you. But the names that brought on the press corps in force were *Agathidium bushi*, *Agathidium cheneyi* and *Agathidium rumsfeldi*, named for President George W. Bush, his vice president and his secretary of defense. Because they were named for beetles that eat slime-molds, cynical colleagues and reporters assumed it was a clever, back-handed slap to the administration. It was not. It was a serious tribute. All the same, the phone rang off the hook. I was living in London at the time and a New York-based news crew flew over to interview me. I did quite a few interviews, especially radio, and managed to dodge repeated calls from one reporter who is known for her mocking man-bites-dog stories.

One day my secretary told me that the White House was on the phone. The story was all over the news, so I figured their press office had some questions. When I picked up the receiver, however, an attractive female voice said, "Professor Wheeler?" "Yes," I replied. "Please hold for the President." During the following seconds of silence my mind raced in disbelief. No … it couldn't be. Was it a prank? The next voice I heard settled the matter as President Bush proceeded to thank me for the honor of having a new species named for him. We spoke for about five minutes and he followed up with a hand written note that makes references to our conversation. I am rather confident that it is the first document penned in the Oval Office that mentions "slime mold."

The potential of DNA sequencing to call out what *might* be new species is being explored, and appropriately so. In general, I believe that formal species descriptions should be based on more than a few differing base pairs. That said, I cannot help but admire the humorous naming of a tiny parasitoid wasp from Thailand, *Aleiodes gaga*, by Donald Quicke and Buntika Areekul Butcher. It is named for the pop singer Lady Gaga. It differs from related wasps by only four bases in the *COI* gene. You guessed it, the sequence of the four is guanine-adenine-guanine-adenine: G-A-G-A! Hours after my column in *The Guardian* reported on this obscure wasp, there were thousands of hits to the newspaper's web site. The potential to reach masses of people through celebrity names is undeniable. While some believe it trivializes or detracts from the dignity of taxonomy's naming of species, I see it as good fun and a lifeline for a science in desperate need of attention.

Making Species Discovery Count

As species descriptions are registered, and up to the minute numbers exist, there should be a web site and billboards in natural history museums that track the discovery of species in real time to add a dynamic sense to the numbers. The Half-Earth Project introduced national report cards so that it is possible to see how well countries are doing in respect to conservation. Similar report cards for the inventory could help the public visualize how much progress is being made, where and for what taxonomic groups. Even before ramping up efforts, there are about 50 new species named every single day. If an inventory increases that by an order of magnitude, a new species will be added on average every three minutes. Such real-time reporting could include total species numbers, as well as subtotals reflecting numbers for various groups of plants or animals. Institutions active in the inventory could tailor readouts for their museum, too, such as the number of new species named by their staff this year, to date and in total.

Taxonomy is disturbingly invisible to the general public and, as Oscar Wilde reminded us "… there is only one thing in the world worse than being talked about, and that is not being talked about." As systematists find their voice, promote their agenda and educate the world about the diversity and origins of life, and biomimicry makes headlines by solving all kinds of problems, we can dare to dream also of a new kind of evolutionary economy that is innovative, adaptive and committed to continual improvement of the prosperity of society and biodiversity.

Further Reading

Allmon, W. D., Pritts, M. P., Marks, P. L., et al. (2017) *Smith Woods: The Environmental History of an Old Growth Forest Remnant in Central New York State*, Paleontological Research Institution, Ithaca, 207 pp.

Anon. (2005) Researchers name beetle after Bush. *Cornell Daily Sun*, 21 April.

Butcher, B. A., Smith, M. A., Sharkey, M. J., and Quicke, D. L. J. (2012) A turbo-taxonomic study of Thai *Aleiodes* (*Aleiodes*) and *Aleiodes* (*Arcaleiodes*) (Hymenoptera: Braconidae: Rogadinae) based largely on COI barcoded specimens, with rapid descriptions of 179 new species. *Zootaxa*, 3457: 1–232.

Fara, P. (2003) Carl Linnaeus: Pictures and propaganda. *Endeavour*, 27: 14–15.

Half-Earth Project (2022) National Report Cards, http://map.half-earthproject.org (Accessed 7 January 2023).

Miller, K. B. and Wheeler, Q. D. (2005) Slime-mold beetles of the genus *Agathidium* Panzer in North and Central America: Coleoptera, Leiodidae. Part II. *Bulletin of the American Museum of Natural History*, 291: 1–167.

Ohr, M. (2018) *The Art of Naming*, MIT Press, Cambridge, 294 pp.

Wheeler, Q. (2012) New to Nature No. 84: *Aleiodes gaga*. *The Guardian*, 16 September.

Wheeler, Q. D. (1987) A new species of *Agathidium* associated with an "epimycetic" slime mold plasmodium on *Pleurotus* fungi (Coleoptera: Leiodidae—Myxomycetes: Physarales—Basidiomycetes: Tricholomataceae). *Coleopterists Bulletin*, 41: 395–403.

Wilde, O. (2007) *The Picture of Dorian Gray*, W. W. Norton, New York, 517 pp.

22 The Evolution of Evolutionary Economics

Capitalism and economic development are frequently spoken about as if they were necessarily antagonistic to the environment. Past experience leads us to associate economic prosperity with excess, waste and negative environmental impacts. Yet, it is only when basic human needs are met that people have the luxury of worrying about environmental issues, much less the fate of species they may never see. Poor nations are not in a position to lead technological innovation designed to lessen impacts on nature. Survival is a stronger instinct than thinking "green." Realistically, we must be concerned with maintaining a vibrant economy if we wish to have the wealth to save biodiversity.

Describing 10 million species or more, an inventory creates the opportunity to revolutionize biomimicry from the supply side. Translating technical taxonomic works, it is possible to populate a database with properties of species described in common language accessible to biomimetic innovators. Searching the phrase "water repellent," for example, could yield hundreds of model species as diverse as *Alchemilla monticola*, whose fuzzy leaves shed water, to the desert beetle *Stenocara gracilipes* that excretes wax to conserve water. With millions of promising clues it is possible to imagine a transformation of the economy into one based on innovation and sustainability rather than exploitation and waste. As knowledge of species and the biosphere grow, so too will options for making things better. Taking full advantage of nature's "knowledge," in large part through an inventory and descriptive taxonomy, we can fundamentally rethink how this externality can enable the economy to innovate, adapt and grow.

The diversity and history of species suggest a different take on evolutionary economics. Comparing itself to microevolution, evolutionary economics has so far focused on process analogies, such as variation, competition and selection. Taking a page from systematics, evolutionary economics can turn to outcomes of evolution for guidance. A double entendre, the version of evolutionary economics that I have in mind refers both to an economy built on mimicking evolutionary novelties and one that is itself ever-evolving in search of new, better, more efficient, less wasteful and increasingly sustainable designs, materials, processes and products. The traditional comparisons with evolutionary processes still apply, but this view emphasizes the results of evolutionary processes—species and characters—as a

DOI: 10.4324/9781003389071-25

conceptual resource. Ideas and models found among species present to us a vast number of options. With specific problems in mind, we act like agents of natural selection favoring models that meet our needs most effectively and efficiently. The goal is to learn as much as we can about the attributes and adaptations of species and their ecological interactions while guarding this source of knowledge for future generations, reducing our footprint on the earth and allowing as many species as possible to persist into the future.

As our understanding of species increases, so too will the number of ideas for improving human industry and quality of life. As systematics shifts into high gear and discoveries flow, low-hanging fruit will appear for designers, engineers and entrepreneurs to harvest. Regardless of the number of successes, perfect efficiency and sustainability are unattainable. Constantly reaching for them, however, we can continually improve step by step. Progress will come in the form of ever-better models, "ideas" derived from species then perfected by human ingenuity to precisely match our requirements.

Species are the product of billions of years of trial-and-error experiments and present to us a mother lode of ideas for reimagining the economy. As biomimetic successes accumulate, we will gain deeper appreciation for species diversity and conservation. Biodiversity and biosphere functions will come to be recognized as the source to which we return, time and again, to fuel the evolving economy.

The mindboggling number of evolutionary success stories should be a source of encouragement as we face what can seem impossible challenges. A taxonomic renaissance, species inventory and biomimetic mindset allow us to imagine an economy that rapidly adapts to overcome challenges and seize opportunities. Comparisons of economies to both evolutionary processes and ecosystem functions make sense at any moment in time, as it encourages us to consider all the factors at play, externalities and all: inputs and outputs, competitive efficiencies, and so forth. The ecosystem analogy in particular helps to conceptualize constraints and interactions, hopefully heightening our appreciation for real ecological services, the most fundamental inputs that keep our society afloat. But current evolution and ecology analogies are not enough to take us where we need to go, into a truly sustainable future. We need inspiration and fresh ideas.

A well-tuned ecosystem in the Permian would be maladapted for prevailing conditions today. And a well-oiled economic system will soon prove unfit as well, as the conditions in which it operates change. Like species, our economy must be constantly aware of its environment, continually experimenting with mutations of the ways we do things. And, like species, each component of our economy must either adapt or go extinct.

Traditional economic theories ignored the value of nature and natural resources. Historically, they were taken for granted, freebies there for the taking. When human populations and economies were small that was a workable approach. No longer. Nature's value proposition is now addressed directly by economists and whole books, journals and international projects are devoted to it. Nature has tremendous economic value, directly and indirectly. It was intellectually dishonest when its value to doing business was ignored. Beyond the value of other species as natural resources, it is time that economists recognize species diversity as an additional value for the ideas it can give us for sustainable living. Acknowledging the biomimetic potential in nature, already happening, will be as transformative to economics as valuing ecosystem services has been. Janine Benyus has pioneered such appreciation of nature as a source of guidance for problem-solving. Her work is a reminder to never discount the value of species or their attributes as an untapped source of drivers for an adaptive economy and a sustainable society.

Fortunately for us, species have been innovating and adapting for a very long time, so we need not retrace their steps. Twenty-four hours a day, year in and year out, for nearly 4 billion years, nature's research and development department has been on the case. The results are as astounding as they are numerous. Millions and millions of species, each a success story in its own right with something to teach us. It is difficult to imagine any problem of significance that nature has not successfully overcome.

Natural selection has bequeathed to us a vast store of successful ideas, designs, materials and processes that can be mined for inspiration to solve the environmental challenges we face today, and those that arise in the future. We need only study and learn from species and their interactions, on scales from species to the biosphere as a whole, to light the path forward. Some ideas and models need only be transplanted from the wild to human systems and practices. Others will require tweaking, but that's okay, too. It is still much faster to be given a promising starting point and improve upon it, even if it requires additional trial-and-error guesswork. Such nature sourced prototypes can vastly decrease the time before solutions can be brought to market.

It is worth noting that even evolution's losers are worthy of study. A much greater number of species have lived and gone extinct than are living today. It is worth remembering that these fossil forms were once well adapted for environments of different eras, in some cases eras that resemble the direction our environment is headed today. Our economy, like evolution itself, should never rest in its quest to detect new challenges on the horizon, adapt and overcome.

An evolutionary economy has the resilience to surmount obstacles that arise, whether they are foreseen or take us by surprise. It is restless, constantly on the lookout for new and better ways to do everything. And just as competition today drives innovation and excellence up, and costs down, fierce competition among species for survival has accelerated beneficial adaptations. Successes in both cases are dictated by powerful forces, whether the market's invisible hand or the heavy hand of natural selection.

Evolutionary economics should regard species diversity as virtually unlimited intellectual venture capital, sitting on the sidelines until needed. Species diversity is a safety net, a source of ideas for new options in the face of unexpected threats. To enjoy an economy benefitting from what we can learn from species, we need, from the outset, to forge some unprecedented partnerships. We must shorten the distances from taxonomic descriptions to ideas that inspire designers and engineers, and from their inspirations to marketable solutions.

Museums and universities should create biomimetic institutes to serve as problem-solving think tanks, similar to what Arizona State University has done, but with the added accelerant of systematics and an expanding database of ideas gleaned from species. Those seeking sustainable alternatives could join a team assembled to address their particular issue. Such teams could consist of taxon specialists, information scientists and designers or engineers as appropriate to survey the landscape of biomimetic models and zero in on the most promising ones. This may mean engaging with a range of botanists, zoologists and microbiologists until the best model is identified, but with millions to choose from a superb model is almost certainly out there. This fusion of taxonomy, information science and hands-on problem-solving is something new, and we will no doubt get better at it as we go. An immediate challenge is to develop software that allows us to search among models based both on phylogenetic relationships and on convergent similarities among unrelated species.

A report prepared for the Australian Academy of Sciences by Deloitte stated that the return on each dollar spent on taxonomy is as high as 35 dollars, supporting the notion that a taxonomic renaissance is a sound investment. The economic promise in biomimicry is massive, even before a revival of descriptive taxonomy. Lebdioui cites a projection that by 2050, in South Korea alone, biomimicry will generate US$382 billion and 2 million new jobs. Just as we awaken to the potential in biomimicry, it would be tragic to permit millions of models to go extinct undocumented. There is nothing we can do to improve the long-term outlook for our economy that is more fundamental or enduring than to unleash the potential in descriptive taxonomy and a species inventory.

As our civilization struggles to overcome challenges, we can do no better than retooling our economy so that it keeps pace with our rapidly changing world. The diversity, adaptations and successes of millions of species provide an incredible wealth of ideas with which to create a brighter future. A worldwide species inventory, done to the high standards of descriptive and phylogenetic systematics, is key to launching a next generation evolutionary economy.

Further Reading

Anon (2013) *Bioinspiration: An Economic Progress Report*, Fermanian Business & Economic Institute, San Diego, 45 pp.

Anon (2018) *Global Biomimetic Technology Market: Focus on Medical & Robotics: (End-User and Application) – Analysis and Forecast, 2018-2028*, BIS Research, Fremont, 138 pp.

Blume, L. E. and Durlauf, S. N., eds. (2006) *The Economy as an Evolving Complex System III: Current Perspectives and Future Directions*, Oxford University Press, New York, 377 pp.

Chami, R., Cosimano, T., Fullenkamp, C., and Nieburg, D. (2022) Toward a nature-based economy. *Frontiers in Climate*, 4: 855803. doi:10.3389/fclim.2022.855803

Conca, J. (2020) How to price the 'natural capital' of planet Earth. *Forbes*, 31 October.

Deloitte (2021) *Cost Benefit Analysis of a Mission to Discover and Document Australia's Species. Prepared for the Australian Academy of Science*. Deloitte Access Economics, Perth, 107 pp.

Diaz, S., Pascual, U., Stenseke, M., et al. (2018) Assessing nature's contributions to people. *Science*, 359: 270–272.

Drupp, M. A., Meya, J. N., Baumgartner, S., and Quaas, M. F. (2018) Economic inequality and the value of Nature. *Ecological Economics*, 150: 340–345.
Friedman, D. (1998) Evolutionary economics goes mainstream: A review of the theory of learning in games. *Journal of Evolutionary Economics*, 8: 423–432.
Hodgson, G. M. (1993) *Economics and Evolution: Bringing Life Back into Economics*. University of Michigan Press, Ann Arbor, 381 pp.
Jing, C. (2016) *The Unity of Science and Economics: A New Foundation of Economic Theory*, Springer, New York, 136 pp.
Johnson, J. A., Ruta, G., Cervigni, R., et al. (2021) *The Economic Case for Nature*, The World Bank, Washington, 157 pp.
Juniper, T. (2012) We must put a price on nature if we are going to save it. *The Guardian*, 10 August.
Kareiva, P., Tallis, H., Ricketts, T. H., Daily, G. C., and Polasky, S., eds. (2011) *Natural Capital: Theory and Practice of Mapping Ecosystem Services*, Oxford University Press, Oxford, 400 pp.
Lange, G.-M., Hamilton, K., Ruta, G., et al. (2011) *The Changing Wealth of Nations: Measuring Sustainable Development in the New Millennium*. World Bank, Washington, 220 pp.
Lebdioui, A. (2022) Nature-inspired innovation policy: Biomimicry as a pathway to leverage biodiversity for economic development. *Ecological Economics*, 202: 107585. doi:10.1016/j.ecolecon.2022.107585
Mateo, N., Nader, W., and Tamayo, G. (2001) Bioprospecting. In *Encyclopedia of Biodiversity*, vol. 1 (ed. S. Levin), Academic Press, New York, pp. 471–488.
Millennium Ecosystem Assessment (2005) *Ecosystems and Human Well-being: Synthesis*. Island Press, Washington, 137 pp.
Mims, C. (2011) The booming business of biomimicry. *Fast Company*, 3 November.
Muradian, R. and Gómez-Baggethun, E. (2021) Beyond ecosystem services and nature's contributions: Is it time to leave utilitarian environmentalism behind? *Ecological Economics*, 185: 1–9.
Naidu, S. G. (2001) Water balance and osmoregulation in *Stenocara gracilipes*, a wax-blooming tenebrionid beetle from the Namib desert. *Journal of Insect Physiology*, 47: 1429–1440.
National Academy of Sciences of the United States (2016) *New Worlds, New Horizons: A Midterm Assessment*, National Academies Press, Washington, 138 pp.
Nelson, R. (2018) *Modern Evolutionary Economics: An Overview*, Cambridge University Press, Cambridge, 272 pp.
Nelson, R. R. and Winter, S. G. (1982) *An Evolutionary Theory of Economic Change*, Harvard University Press, Cambridge, 454 pp.
Ouyang, Z., Song, C., Zheng, H., et al. (2020) Using gross ecosystem product (GEP) to value nature in decision making. *Proceedings of the National Academy of Sciences*, 117: 14593–14601. doi:10.1073/pnas.1911439117
Polski, M. (2005) The institutional economics of biodiversity, biological materials, and bioprospecting. *Ecological Economics*, 53: 543–557.
Reid, W. V., Laird, S. A., Meyer, C. A., et al. (1993) *Biodiversity Prospecting: Using Genetic Resources for Sustainable Development*, World Resources Institute, Washington, 341 pp.
Shiozawa, Y. (2004) Evolutionary economics in the 21st century: A manifesto. *Evolutionary and Institutional Economics Review*, 1: 5–47.
Swanson, T. (1996) The reliance of northern economies on southern biodiversity: biodiversity as information. *Ecological Economics*, 17: 1–8.

Epilogue

What we have today is biodiversity bedlam: species going extinct randomly by the thousands; efforts to conserve biodiversity without a clearly defined end-game; dependence on ecological services from "dark biodiversity," millions of species that we do not know; a desire to replace wasteful and environmentally destructive practices, but few viable alternatives; a deep yearning to understand our selves and world, about to be dashed by the destruction of evidence of evolutionary history; DNA-based glimpses of species, when it is deep knowledge that we need. We are in an uproar about the environment, extinction and climate change. But our responses are ineffective due to lack of knowledge, context and clearly articulated goals. It is irrational and self-defeating to remain ignorant of species in a period of mass extinction and profound planetary change.

It is time for a taxonomic renaissance. Time to invest in the expertise and infrastructure necessary to complete an inventory of earth's species. No mere list of names, or database of DNA sequences, an inventory must be an information-rich account of all species, organized in a predictive, phylogenetic classification. It is time to remember why the science of systematics came about in the first place: to address our curiosity about the pattern of similarities and differences among species, and its origins. That curiosity led to species discoveries that, in turn, made possible agriculture, antibiotics and a theory of evolution, to name three examples among many. And, it is time to raise expectations for natural history museums as permanent records of species diversity.

Reviving descriptive taxonomy is important. Descriptions document the most interesting, complex and surprising outcomes of evolution, precisely the things we hope to better understand by reconstructing phylogenetic patterns. They detail what makes each species unique, often making it possible to identify species on sight in the field where natural history observations may be made. Morphology characters link living and fossil species and are the things we hope to explain with embryology and genomics. Knowledge of such structures often makes it possible to infer their functions. And, descriptions bring biomimetic models to light.

Everyone should have the epiphany about systematics that Janice Pariat experienced researching her novel *Everything the Light Touches*. She set out to portray Linnaeus as "intent on categorizing and labeling, on reducing the natural world into an immutable, mechanistic collection of parts." Instead,

> … the 25-year-old Linnaeus I encountered … was filled with wonder. It was powerful, and undeniable—his magnificent love for the natural world. The awe he felt, the affection, the careful attention he bestowed upon the living beings he came across—including the gnats that vexed him so!

Perhaps she mistook Linnaean taxonomy for DNA barcoding.

We expect physicists and astronomers to dream big, conceive grand projects and design instruments to carry them out. "Big science" projects are increasingly part of biology, such as the Long-Term Ecological Research Network and National Ecological Observatory Network in the U.S., and Brazilian Flora 2020, an inspiring step toward a comprehensive world flora. In 2020, the Earth BioGenome Project set out to sequence the approximately 2 million named eukaryotic species in ten years. Such bold thinking is exactly what is called for on a planet whose biodiversity and ecosystems are in jeopardy. However, even if the latter project were to succeed, it would tell us very little about species. Two million named eukaryotes represent a small fraction of earth's species and as we sequence them, others are going extinct.

There is an assumption that taxonomic revisions and monographs, based on morphology, are a thing of the past. This is part of the misguided belief that we need a new system based on genomic information, a systematics that looks more like genetics. But replacing existing taxonomy with a molecular approach in the middle of an extinction crisis is like translating scrolls at Alexandria as the library is consumed in flames.[1] Genomic data is a potent addition to taxonomy, but hardly up to the task of replacing morphology or fossils for understanding evolutionary history or species diversity. We must support a far more ambitious project, a fusion of the best of traditional taxonomy with modern technologies. A project that integrates all relevant evidence, tests species and phylogenetic hypotheses, creates and improves classifications—all while ferreting out, describing and classifying the eight or more million species that remain entirely unknown.

Born in the Age of Enlightenment, modern taxonomy was the original biological "big science" project. Linnaeus' vision of discovering and classifying all species is now seen as a moonshot-scale ambition. To explore our planet's biodiversity, we don't need to conceive an entirely new project, replace everything we have accomplished since Linnaeus or kiss the ring of pop science. We simply need to finish his campaign of discovery that has been underway for more than two and a half centuries. What has changed since the time of Linnaeus is the feasibility of completing such a planetary-scale enterprise. Impossible in his day, we have now the means of travel and communication, and the theories, methods and tools, with which to efficiently discover, describe, name and classify every kind of plant, animal and microbe on our planet. We have seen Linnaeus' project as impossibly big for so long that we overlook the plain truth in front of us. We can achieve what generations before us could not: a comprehensive, worldwide account of life on earth.

Some have argued that professional taxonomists are a thing of the past. For them, the future consists of molecular labs, mothballed collections and punting the tasks of collecting and describing species to a rag-tag army of amateurs and part-timers. Taxonomy is an unusually inviting place for specially trained amateurs. But at its heart, systematics remains a very serious, intellectually challenging and professionally demanding field. Amateur contributions should be encouraged, welcomed and vetted, but the driving force for advancing knowledge in the future must be led by a community of doctoral-level researchers, scholars and curators. Many theoretical and on-the-ground challenges remain for a science that only entered its most recent incarnation—as a discipline capable of producing rigorously testable, phylogenetic classifications—little more than 50 years ago. By any objective measure, the traditional mission of systematics is more demanding and rewarding than lab procedures that can be carried out by trained technicians. Only a superficial view of science would see the whistles-and-bells of molecular labs as a worthy successor to the deep scholarship, diverse information and rigorous theories of traditional systematics.

We cannot afford to accept a substitute for serious systematics. Not now. Not ever. And especially not during a mass extinction. Genomic sirens with promises of quick, effortless species identifications lure us off course into perilous waters. After centuries of species exploration and classification it is wrongheaded to consider, for even a minute, alternatives that result in less, and less reliable, information. An undeniable aspect of the ongoing extinction of species is that the time for a comprehensive inventory is short. In a matter of decades, we are challenged to discover and describe at least five times as many species as we have done since 1758. Happily, it is not too late to save and modernize systematics, efficiently complete an inventory, inform conservation decisions and mine a vast deposit of biomimetic models.

Systematists must not sell out the integrity of their science for short-term funding or the approval of colleagues who neither understand nor appreciate their mission. They must not permit taxonomy to be redefined in the image of, or as a service to, other fields. Because we have exactly one chance to inventory and document species, it is imperative that taxonomy be done to its highest standards of excellence, without compromise or distraction. It is easy to see our situation as hopeless living at a time of rampant extinction and looming environmental threats. But that is wrong. Not because we are not witnessing gross degradation of the natural world, but because in prematurely accepting defeat we ignore fantastic opportunities.

We alone have the chance to discover and document the diversity of species as they exist at the dawn of the Anthropocene. Only we can gather and preserve evidence of evolutionary history otherwise lost. And we can, if we are wise, create museum collections and a great body of knowledge with which we can continue to learn about evolution, maintain a high standard of living and protect a biosphere inhabited by diverse plants and animals. Future generations will only dream of such opportunities. For us, we need only act rather than despair.

Linnaeus' enterprise is in tatters. Museums have lost the plot. Retiring taxonomists are not replaced in kind. Descriptive taxonomy has been de-funded. Metagenomic environmental surveys and metrics of genetic distances are not recognizable as taxonomy to me, and I doubt they would be to Linnaeus, Darwin or Hennig. Yet, all the ingredients for a resurgence of the systematics agenda are in place. Theoretical obstacles have been largely overcome. Technologies exist to speed progress. A number of bright students are attracted to the intellectual challenges and rewards of taxonomy in every generation. And, centuries of taxonomic progress have laid the foundation for a worldwide inventory.

Systematics is a science like no other. As exciting as astronomy, audacious as cosmology and as impactful as biomedicine or agriculture, it engages us in a personal way. What other science takes you on a journey as wide as the earth, as deep as time and as diverse as life itself? What greater adventure could there be than exploring life on a little-known planet? What understanding is more rewarding than an overview of the origin and diversification of biodiversity? Beyond personal relationships, what is more intimate than knowing other beings in minute detail, or sharing great ideas with colleagues? There is no satisfaction greater than understanding the totality and history of a lineage, captured in names, descriptions and classifications. What other science so seamlessly combines rigorous theories with immersive, childlike wonder? And, what could be more human than honoring non-humans with the dignity of being recognized?

After decades of neglecting systematics, the time has arrived to return it to its rightful, center-stage role in confronting the biodiversity crisis. We can no longer permit greed, hubris and ignorance to deny support to the one science capable of documenting the diversity and history of life. I know no taxonomist who would deny prestige or funding to any other field of science, whether physics, chemistry or molecular genetics. But reciprocal respect has not existed for decades. This lack of generosity of spirit cannot be permitted to cause us to miss the fleeting opportunity to complete an inventory of species.

As a rule, scientists are intelligent … but they're not always wise. Science is an amazing human activity with a wondrous capacity to recognize and fix its own errors. But such course corrections can take a generation or more to effect, and time is running out for many species. There are times when public opinion has an important role to play, returning common sense to the process of setting priorities in both politics and science. Banning DDT, in the wake of Rachel Carson's *Silent Spring*, was one such time. Today is another. Species and their improbable evolutionary story are our collective heritage, and the public has the right—I would say responsibility—to demand, alongside professional taxonomists, that funding be restored to systematics. Mobilizing such support will depend on a greater number of taxonomists becoming effective ambassadors for their science. Students should take it upon themselves to consciously hone communication skills and situational awareness, being on the watch for discoveries ripe for dissemination to the popular press.

Some of my taxonomic colleagues are so demoralized that they have given up, accepting the loss of most species as inevitable, and surrendering systematics to molecular methods and servitude. Clear-eyed about what is happening to species and systematics, I choose optimism. Taking decisive action, we can mitigate the loss of species and gain an enormous amount of knowledge, even from those species that ultimately succumb to extinction. The extent to which systematists complete an inventory and provide the knowledge needed for conservation and biomimicry is yet to be seen, but it will be far greater if we approach species exploration with the urgency it deserves. I am grateful that we have recognized the extinction crisis before huge numbers of species were gone. It is not too late to conserve millions of species … and preserve evidence that millions of others existed.

This book is a plea to revive descriptive taxonomy, expand and develop natural history collections, complete a species inventory, lead a biomimetic revolution, establish a truly evolutionary economy and conserve as many and as diverse species as we can. If we begin today, we can have a first-pass inventory of species in as few as 50 years, changing the course of history for people, science and the planet. And we will have positioned systematics to continue far into the future, deepening our knowledge and appreciation of species and evolution, and expanding options for living sustainably.

In order to save systematics, it is imperative that we not lose sight of its mission and goals. I am critical of molecular approaches to systematics when they focus on merely telling species apart and are disproportionately favored. Not because there is anything inherently bad about molecular data. It is one of several important sources of evidence but, like all data, has strengths and limitations. Along with comparative morphology, paleontology and embryology, molecular data, properly interpreted, reflects the pattern forged by evolutionary history. Many DNA analyses today, however, are phenetic. That is, they assess the overall similarity of sequences rather than teasing out discrete, shared-derived characters (i.e., synapomorphies) suited to interpreting historical patterns. This, of course, refers merely to data gathering and analysis and is easily fixable. What really bothers me is something more fundamental in the current molecular movement: it has a different end in mind.

The aim of molecular-based taxonomy is to tell species apart, estimate their degree of relatedness and provide identifications to other biologists. Those are worthy aims as far as they go, but they are quite different from those of traditional systematics. Systematists seek to understand species as individual, unique kinds of organisms. That is possible only when the focus is on species and their attributes, not just their DNA. This, of course, requires that we carefully compare, describe and analyze their characters. This divergent goal is why molecular studies often publish branching diagrams showing relationships among species from which all interesting characters have been omitted. Such studies simply recognize species and their relationships without ever knowing what makes them interesting.

Returning to an astronomy analogy, it is possible to tell planets apart by their position in space. Earth is the third planet from the star known as the Sun, and so forth. With a little more knowledge, we can understand the effects of gravitation on their orbits and learn the average number of planets in star systems, among other aspects of the organization and workings of the Universe. But such superficial understanding robs us of the most fascinating details of planetary science. Describing whether a planet is solid or gaseous, large or small, the chemical composition of its atmosphere, its surface temperature and other physical attributes make studying planets far more rewarding. Knowing that Neptune has 14 moons and earth has only 1, or that Saturn has rings and Mars does not, tells us fascinating things and raises interesting questions. Simply telling planets apart, being able to identify them based on spatial relationships, while learning nothing about what makes each unique, is a rather unambitious enterprise. Yet, this is what is on offer from a molecular-based taxonomy.

Systematics is all about the diverse attributes of species and groups of species, and the patterns among them due to a shared history. Species and characters open two views on the same historical sequence of events. You cannot be truly interested in species themselves without the irresistible urge to describe, compare and classify them, yet that is precisely what molecular systematists are doing because they have different goals.

Once you change the ends, it becomes easy to justify shortcuts in the means. It is possible to meet the limited needs of other biologists to identify species and to get an idea of relationships without detailed, comparative studies or expansive museum collections. Only by retaining the traditional mission and goals of systematics can we have both. Systematists can explore the characters and history of species, spinning off the knowledge and classifications with which other biologists can accurately identify species and see them in their phylogenetic context.

We must not allow systematics to be redefined because it makes the minimal information sought by others cheaper or easier to get, not when it comes at the cost of never understanding species diversity and history. Supporting molecular shortcuts to meet needs of users of taxonomic information in exchange for the rigor and intellectual depth of systematics should be an unacceptable bargain. Only by supporting systematists to pursue their own science can we fully meet the needs of both taxonomists and users of taxonomic knowledge. For the systematist, species, characters, cladograms and classifications represent carefully formulated, rigorously testable hypotheses. This means information made available to users is better, too. On a planet on which species are hurdling toward extinction, this is surely preferable to simply telling a rapidly diminishing number of species apart.

Systematics deserves to resume its centuries-old quest to explore and understand the diversity of species, ultimately expressed in detailed, comparative, descriptive studies and classifications that synthesize and organize what we know in a system of information-rich groupings and names. Systematics deserves our respect and support as an independent science, as well as a source of information necessary to ecology, conservation, biomimicry and society. Best of all, in the process of saving systematics, we have the opportunity to save, too, a greater number and diversity of living species, as well as preserving evidence of all species for the enlightenment of future generations.

If we let the last opportunity to discover, describe and know species pass us by, we diminish our humanity, too. We are intellectually impoverished by each detail permitted to disappear. Callous indifference to other species, not even caring enough about them to recognize their uniqueness and give them names, eats away at our ethics. As sentient beings reshaping the biosphere, we owe it to ourselves to recognize and honor our origins and relationship to other species. Ignorance of species widens the gap between people and the natural world, denying our innate love for, and curiosity about, the world around us. And permitting species to disappear without discovering what they can teach us about survival and sustainable living unnecessarily places our future welfare at risk.

The choice could not be clearer. We can continue down the road we are on, ignoring systematics, trading rigorous hypotheses for expedient molecular methods and settling for simply telling species apart. Or, we can pursue fundamental systematics to understand the living Cosmos and how it came to be. It is no longer acceptable to approach science as a popularity contest when so much is at stake. It is time for a sober reprioritization of the roles of taxonomy and natural history museums. There will be no second chances to redress missed opportunities. No other worlds can tell us about earth's history. Only saving systematics can seize this one-time, not to be repeated, opportunity to explore species with direct consequences for our well-being and intellectual fulfillment.

We will never know how many species have gone extinct on our watch, unknown, unnamed and unsung. But this hemorrhaging of biodiversity and knowledge need not continue. It is time for taxonomic triage, an all-out effort to discover and classify every kind of living thing. Knowledge of species will make us better stewards of the planet and more adept at adapting to environmental change. The humility we gain by honoring millions of species with recognition will make us more humane beings, too. It is clearly in the best interest of humankind and the planet to conserve as many, and as diverse, species as we can. It is equally clear that a large number of species will go extinct in spite of heroic conservation efforts. We should not compound the tragedy of extinction by permitting species to disappear without preserving more evidence of their existence than a bit of DNA. Reviving systematics, we can rapidly and efficiently complete an inventory, description and classification of earth's species.

Note

1 I realize that the Great Library at Alexandria is thought to have decayed over centuries rather than burning all at once, but this mythical, cataclysmic version of the story is so much better for my purposes.

Further Reading

Blaxter, M., Archibald, J. M., Childers, A. K., et al. (2022) Why sequence all eukaryotes? *PNAS*, 119: e2115636118. https://doi.org/10.1073/pnas.2115636118

Brazil Flora Group (2021) Brazilian Flora 2020: Leveraging the power of a collaborative scientific network. *Taxon*, 71: 178–198.

Brazil Flora Group (2022). Brazilian Flora 2020 project – Projeto Flora do Brasil 2020. Version 393.354. Instituto de Pesquisas Jardim Botanico do Rio de Janeiro. Checklist dataset, doi:10.15468/1mtkaw (Accessed via GBIF.org, 20 December 2022).

Pariat, J. (2022) Chasing Linnaeus. *The Linnaean*, 38: 37.
Potapov, A., Bellini, B., Chown, S. L., et al. (2020) Towards a global synthesis of Collembola knowledge:—Challenges and potential solutions. *Soil Organisms*, 92: 161–188.
Wilson, E. O. (2016) *Half-Earth: Our Planet's Fight for Life*, Liveright, New York, 259 pp.

Glossary

AI Artificial intelligence. Computer systems performing tasks usually associated with human intelligence.
Apomorphy A derived character.
Aristotle (384–322 BC) Father of biological classification.
Artificial classification Classification based on arbitrary criteria.
Barcoding See DNA barcoding.
Biodiversity As generally used, any and all diversity found in and among living things, from molecules to ecosystems. In this book, the diversity of life *at and above* the species level.
Biogeography Study of the geographic distribution of species.
Biomimetics See *biomimicry*.
Biomimicry Mimicking species or phenomena in nature to solve problems.
Character (pl. characters) Heritable attribute constantly distributed within a species or group of related species.
Character state One of two or more alternative expressions of a character.
Clade A branch in tree-like diagrams depicting relationships among species.
Cladistics Study of the pattern of relationships among species or higher taxa (cf. Phylogenetics).
Cladogram A branching diagram that depicts the relative pattern of relationships among species or groups of species.
Classification The delimitation, hierarchic ordering and ranking of taxa.
COI See *cytochrome*.
Cytochrome c oxidase subunit I Mitochondrial gene used in DNA barcoding to identify animal species.
Darwin, Charles Author of *On the Origin of Species* and theory of evolution by natural selection.
Descriptive taxonomy See *taxonomic revision* and *monograph*.
DNA Deoxyribonucleic acid.
DNA barcoding Use of short segment of DNA to identify species.
Evolution, as noun As used here, the history of life reconstructed as a *pattern* of relationships among species and clades (cf. evolution, as verb).

Evolution, as verb The *process* or *processes* by which species change through time and new species/characters arise, including mutation and natural selection (cf. evolution, as noun).
Evolutionary biology The study of evolution.
Evolutionary economics An economy adapting to changing circumstances and/or using attributes of species as models for biomimicry.
Experimentalism Reliance on experimental procedures.
Extinction Theoretically, the death of the last individual representing a species globally, a population locally, or one of two or more alternative states of a trait within a species.
Falsificationism Belief that nothing can be proven, only shown to be false by additional observations or experiments.
Genome Complete set of genes or genetic material.
Hennig, Willi (1913–1976) Father of phylogenetic systematics.
Homology Identity among attributes due to shared ancestry.
Hotspot (biodiversity) A geographic area with especially large numbers of species.
Hypothesis A claim or generalization.
Kardashians An American family famous for being famous.
Linnaeus (Carl von Linné) (1707–1778) Father of modern taxonomy.
Macroevolution Study of pattern of relationships at and above the species level, or of processes affecting it.
Metabarcoding DNA (or RNA) barcoding method allowing for simultaneous identification of many species in one sample.
Metagenomics Study of entire nucleotide sequences isolated from all organisms in a bulk sample.
Microevolution Study of evolutionary processes within species or populations.
Molecular systematics Systematics based primarily or exclusively on molecular data.
Monograph An exhaustive treatment of a higher taxon (cf. Revision).
Monophyletic group A group including an ancestral species and all (and only) its descendant species.
Morphology Study of form, structure and relationship among parts of an organism.
Natural classification A classification that reflects phylogenetic relationships.
Natural selection A process or processes tending to favor the survival and reproduction of organisms with one or more traits.
Nomenclature The scientific practice of naming species and groups of species.
OTU Operational taxonomic unit; a provisional grouping of similar individuals.
PCR See polymerase chain reaction.
Phenetics Grouping species based on overall similarity.
Phenome What an organism looks like; sum of its traits and characters.
Phylogenetic analysis See *phylogenetics*.
Phylogenetic classification A classification reflecting pattern of relationships among species and groups of species.

Phylogenetic diversity Number of clades represented among species or higher taxa.

Phylogenetic tree In theory, a tree-like diagram depicting actual ancestor-descendant relationships among species. Unknowable in practice, the term is informally applied to cladograms depicting relative relationships and referring only to hypothetical, unnamed ancestor species.

Phylogenetics Study of relative relationships among species and higher taxa based on shared-derived characters.

Phylogeny (pl. phylogenies) The branching sequence of species through time; see *phylogenetic tree*.

Polymerase chain reaction Technique to amplify segment of DNA by producing millions of copies.

Polymorphism Two or more alternative states of a trait, such as blue and brown eye color in humans.

Ranks Categories in a Linnaean classification; e.g., genus, family, order.

Revision Review of a higher taxon incorporating new material and interpretations, typically including identification keys, diagnoses and descriptions (cf. Monograph).

Scala naturae Aristotle's idea that species can be arranged in ladder-like fashion according to their degree of perfection with humans on the top rung.

Science Claims, generalizations or predictions that can be objectively tested by observation or experiment.

Sister species Two species that are each other's closest relative.

Speciation The process(es) by which an ancestral species gives rise to new species.

Species (pl. species) Kinds of organisms; characterized by a unique combination of characters and reproducing to give rise to more of like kind.

Synapomorphy (pl. synapomorphies) A shared-derived character.

Systematic biology See *systematics*.

Systematics The comparative study, description, naming and classification of species and higher taxa, including analysis of cladistic (phylogenetic) relationships. Used here as synonym for taxonomy.

Taxon (pl. taxa) A species or group of species that has been given a name, such as a genus or family.

Taxonomic revision See *revision*.

Taxonomy As used here, synonym for systematics.

Tokogenetic relationships Relationship between parent and offspring, producing reticulate patterns within sexually reproducing species.

Trait (pl. traits) Heritable attributes that vary within or among populations of a species, e.g., hair and eye color in humans.

Tree of life See *phylogeny*, *cladogram*.

Index

Note: Page numbers in *italic* refers to Tables.

Abies alba 112
abundance, species 74, 79, 133, 134, 200
acarologist 32
Acer nigrum 29; *A. saccharum* 29
acid rain 17
Acorn, John 133
Adirondack Mountains 17
Africa 89, 162, 171, 206
Agassiz, Louis 98
Agathidium 56, 144, 145, 218; *A. aristerium* 215; *A. bushi* 218; *A. cheneyi* 218; *A. pulchrum* 56; *A. rumsfeldi* 218
Agra 58, 122; *A. cadabra* 58; *A. phobia* 58; *A. vation* 58
AI *see* artificial intelligence
Alchemilla monticola 220
Aleiodes gaga 218
Alexander, Charles Paul 49
algae 32, 134, 172; blue-green 8
Ali, M. Ajmal 3
amateur 54, 69, 106, 154, 173, 181
amazon 57, 79, 83, 85, 132, 207
American Museum of Natural History 49, 85, 110, 178, 201, 202
amphibians 75, 79, 132, 133, 185
Anax junius 61
angiosperms 65, 133
animals 3, 17, 22, 23, 30, 32, 42, 48, 68, 80, 83, 84, 108, 124, 130, 152, 219, 227; beauty 51; cancer cure 163; comfort 153; cruelty toward 38; cute 38; endemic 132; extinction 132, 133, 156; invertebrate 133; legless 65; number 76, 129; registry of names 169, 174; quadruped 65; taxidermy 205; vertebrate 65, 79, 129, 146, 154, 184
Anopheles 83
Antarctica 77
anthropocene 203, 227
anthropocentrism 36
anthropology 53, 121
Apalachicola River 163
Apophyllus now 58
Appalachian Mountains 135, 145
aquifers 77
Arabidopsis 94
arachnids 39
arachnologist 32
Araneae 66
Araneidae 90
Archaea 8
Archaeopteryx 94
Aristotle 30
Arizona State University 56, 223
art 36; of survival 23; works of 36, 54
arthropoda 76, 79, 113, 133, 154, 206
artificial intelligence 175, 178
arts 50
astronomy 26, 28, 63, 91, 184, 186, 187, 228, 229
astrophysics 30, 184
ATBI (All-Taxon Biodiversity Inventory) 171, 172
athens 88
Atlantic Forest 132
Atractocerus 55
Attenborough, David Frederick 74
Australia 89, 191, 223
Australian Academy of Science 191
Australopithecus africanus 36
Aves 36, 113; *see also* birds

Ba humbug 58
bacteria 8, 57, 134

Bailey, Liberty Hyde 156, 209
Ball, George 55
Banks, Joseph 70
barcode 93; DNA 4, 5, 82, 98, 108, 111, 145, 175
Barnard, Peter 58
Barro Colorado Island 108
Barrowclough, George F. 85
Bartram, John 130
Bartram, William 130
beauty 38, 51, 123, 143, 156
beech tree 132
beetles 29, 36, 38, 53, 55, 57, 58, 79, 80, 85, 121, 122, 131, 135, 144, 154, 171, 184, 189, 207, 215, 220
Benyus, Janine 23, 111, 160, 162, 222
Berlin 189
Berra, Yogi 119, 154
big data 4, 28
big science 19, 31, 187, 226
binomial 112
biodiversity 17, 23, 34, 36, 37, 38, 42, 71, 74, 78, 225; abundance 79; baseline 106; conservation 18, 35, 41, 108, 118, 134, 195, 212, 216; crisis 12, 18, 59, 65, 71, 91, 93, 111, 121, 140, 168; dark 27, 31; distribution 89; DNA 18, 49; economy 221; experience 152; extraterrestrial 40; hotspots 83; language of 111; memory 78, 106; mining 160; monitoring 3, 31, 78, 90, 104, 110, 147, 168; picture of 74, 78, 79; public face 198; survey 183; value 149
biogeography 89, 90, 106, 143, 183, 216, 217; vicariance 89, 188
Biologia Central-Americana 95
biological species concept 85
bioluminescence 97, 170, 171
biomass 79
biomedicine 62, 147, 163, 228
biomimicry 23, 34, 42, 90, 111, 130, 135, 155, 160–6, 168, 172, 195, 211, 220, 229
biophilia 108, 153, 155
biosphere 12, 15, 17, 31, 34, 39, 42, 77, 85, 90, 107, 109, 153, 160, 203, 221
birds 14, 18, 22, 23, 32, 79, 80, 85, 110, 113, 132, 162, 172, 189; *see also* Aves
blue book 47
books, smell 102; value 95–97
Boorstin, Daniel J. 82
Borror, Donald J. 61
botanical gardens 3, 17, 93, 120, 149, 190, 207, 217
Brazil 79, 132, 170, 185, 226
Brazilian Flora 2020 226
Bremer, Kåre 64
Brewster, Benjamin 45
Brooks, Steve 90
Brower, Andrew V. Z. 11
Brown-Wing, Katherine 79
Browning, William 156
bryologist 32
Buffon, Georges-Louis Leclerc, Comte de 70
Burgess shale formation 88
Bush, George W. 218
Butcher, Buntika Areekul 218
butterflies 32, 51, 133, 183, 184, 216

capitalism 39, 58, 220
Carabidae 55, 57
carrying capacity 160
Carson, Rachel 228
Casey, Thomas Lincoln 57
caves 77, 98, 132
Cavia porcellus 144
cell, largest 51
centers of excellence 134, 172, 173, 178, 188, 189, 193, 194, 198, 205, 212
Central America 85, 95, 144, 145, 189
Chagas disease 144
Chapman, Arthur David 88
character 15, 65, 145, 165, 176, 182, 190, 200, 220, 229, 230; diagnostic 174; divergence 122; fixed 66; morphological 68, 141, 225; transformation 71, 89, 141
chemistry 49, 71, 94, 96, 228
Chiapas 55
chironomidae 90
chiropterologist 32
chronology 52, 53, 56, 88, 99, 175
ciliates 155
Chrysanthemum 29; *C. indicum* 30
Chupacabra 104
citizen science 165, 173, 176, 189, 190, 194, 211
civilization 1, 12, 23, 37, 88, 111, 161, 169, 199, 203, 223
clade 14, 66, 78, 82, 87, 147, 148
cladogram 14, 74, 169, 230
classification 1, 4, 8, 11, 14, 18, 26, 28, 42, 45, 47, 58, 63, 65, 74, 106, 141, 148, 165, 169, 175, 182, 188, 226; as general reference system 50, 113; phylogenetic 11, 19, 25, 28, 67, 93, 113, 163, 205, 225; predictive 19, 90, 163; climate change 8, 34, 41, 47, 90, 107, 118, 127, 130, 134, 225

Coddington, Jonathan 58
code of conduct 124
Cognato, Anthony 3
collecting 48, 48, 53, 55, 71, 119, 121, 154, 170, 189, 192–5, 201
collections, natural history 3, 10, 17, 30, 44, 69, 80, 82, 92, 105, 140, 154, 162, 167, 177, 181, 189, 196–12, 226, 227
colombia 185
comparative morphology *see* morphology
Compositae 113
computer assisted tomography 100
Comstock, Anna Botsford 158
Congo 83
Congress of the United States 42, 62, 186
conservation 5, 12, 18, 20, 34, 40, 78, 135, 153, 172, 203, 208, 212, 217, 221, 231; *vs*. collecting 192; commitment to 34, 153; ethic 152, 155; *ethos* 18, 168; goal of 118–25, 169; human-centric 37; measurable 18, 89, 104, 107, 109, 110, 147, 203; policy 181; tree-hugger 121
continental drift 89
Convention on Biological Diversity 193
Copernicus 47
Cornell University 47, 209; Press 156
Corylophidae 58
cosmology 26–33, 52, 228
Cowie, Robert H. 130
Cracraft, Joel 84
Creationism 52, 71, 112
creative writing 54
Cretaceous 53, 135
Cretaceous-Tertiary boundary event 135
Critterpedia 178
crocodiles 142
Croizat, Leon 188
Crowder, William 51
Crowson, Roy 56
CSIRO (Commonwealth Scientific and Industrial Research Organisation) 178
Culicidae 8
curation 70, 143, 205, 210
Curculionoidea 75
curiosity cabinets 205
Curtin, Jane 46
Cuvier, Georges 70
cyber-enabled taxonomy *see* cybertaxonomy
cyberinfrastructure 90, 106, 168, 185, 187, 210
cybertaxonomy 19, 168, 170, 187, 194
Cyprinus rubrofuscus 8

dark age 31; dark biodiversity 27, 31; dark energy 27; dark matter 27
Darth Vader 218
Darwin, Charles 22, 40, 47, 65, 70, 88, 99, 100, 113, 146, 217, 228
Darwinism 112
Dasycerus maculatus 135
Dawkins, Richard 89
DDT (dichlorodiphenyltrichloroethane) 228
Dead Zoo *see* Natural History Museum, Dublin
DeLong, Dwight M. 49
demarcation principle 46
Denmark 133
descriptions, taxonomic 67, 146, 147, 162, 174, 178, 187, 195, 199, 211, 218, 222, 225
developmental pathways *see* ontogeny
Devine design 146
Diamond, Jared 85
diatoms 77, 132, 175, 206
Dickens, Charles 169
digital imaging 19, 69, 100, 173, 174, 175, 190, 210
digital technologies 19, 168, 170, 171, 177, 178, 187, 193, 206, 211
Dijkstra, Klaas-Douwe B. 143
dinosaurs 36, 53, 130, 189, 203, 206, 210
Dipterist 32
Distributed European School of Taxonomy 190
DNA data 3–6, 9, 11, 15, 17, 18, 20, 28, 30, 35, 40, 48, 51, 66, 71, 77, 82, 87, 91, 92, 94, 95–101, 105, 111, 113–15, 141–9, 162, 173, 192, 199, 201, 218, 225, 226, 229, 231; boring 49; information content 18, 69; revelatory 68; superiority 68
Dobzhansky, Theodosius 71
Dollo, Louis 101
Dollo's law 101
dragonfly 61

e-monograph 145, 174–177, 187, 188, 195; *see also* monograph
Earth BioGenome Project 226
Ebach, Malte xii
Eberhard, William 58
ecological services 12, 18, 35, 107, 111, 123, 149, 221, 225
ecology 3, 5, 18, 19, 74, 83, 106–9, 123, 149, 160, 171, 200, 203, 208, 212, 214, 217, 221, 230

economics 41, 123; evolutionary 220–3
ecosystem 8, 10, 14, 17, 18, 23, 26, 31, 34, 35, 36, 38, 44, 78, 87, 104, 106–15, 119, 122–4, 132, 133, 147, 152, 161, 168, 221
ecosystem science 18, 36, 217
Edison, Thomas 24
education, biologists 62
Egypt 46
Ehrlich, Anne H. 130
Ehrlich, Paul 130, 160
Einstein, Albert 3
embryology *see* ontogeny
emergent properties 71, 72, 94
Emerson, Ralph Waldo 63
emotion 18, 21, 22, 46, 51, 93, 119, 156, 157, 208
encyclopedia of life 33
endangered species 37, 119, 123, 152
Endangered Species Act 119
endemism 132
Engel, Michael 76
English, rules of 55
environment 37, 39, 77, 82, 87, 94, 98, 105–7, 119, 192, 211, 220, 225; artificial 120; baseline 78; built 152, 156; change 8, 12, 14, 34, 36, 38, 41, 78, 91, 107, 124, 135, 203, 231; challenges 18, 19, 23, 24, 42, 90, 104, 106, 127, 133, 134, 140, 147, 157, 160, 161, 194, 198, 203, 222, 227; decisions 17; human-made 39; stewardship 34
epigenetics 94
epistemology 52
Equisetales 37
Equisetum 37
Erwin, Terry L. 58, 76, 110
Escherichia coli 77
ethnobiology 111
Euglena gracilis 155
eukaryotes 226
Europe 46, 75, 83, 112, 135, 143, 186, 190, 193
European Science Foundation 186
Evans, Howard Ensign 26
Evenhuis, Neil Luit 58, 191
evolution 6, 10, 11, 22, 26, 31, 32, 34, 35, 37, 39, 42, 44, 69, 76, 92, 93, 108, 110, 11–15, 129, 143, 146, 148, 152, 158, 214; chance 88; convergent 56, 145, 163; economic 220–3; history 12, 14–18, 28, 30, 38, 40, 71, 82–4, 87–91, 97, 105, 109, 113, 130–5, 142, 168, 169, 182, 184, 198, 202, 204; laws 71, 101; macro 78, 140; micro 140; neutral 3; novelties 11, 68, 74, 86, 141; patterns 67, 85, 140; process(es) 52, 66, 72, 79, 84, 85, 87
exobiology 27, 43, 105
experimentalism 10, 12, 25, 26, 39, 52, 64, 65, 79, 93, 109, 114, 121, 139, 140, 191, 201, 210, 214; non-experimental 44, 52, 64, 78, 108, 140; universe of outcomes 64
extinction 4, 8, 12, 17, 18, 20, 24, 25, 31, 35–7, 41, 43, 63, 71, 80, 90, 104, 106–8, 110, 118–24, 129–35, 143, 147, 152–6, 165, 178, 198–202, 225, 227, 231; background rate 130, 132; engine of evolution 37; mass 11, 12, 34, 38, 70, 140, 157, 167, 192, 211, 214, 216, 226
extraterrestrial life 104, 105, 155

fads 12, 17, 19, 54, 104, 143, 178, 179, 208, 209, 214
Fairmaire, Léon Marc Herminie 53
Fall, Henry Clinton 144
falsificationism 64
Fawcett, Frances 75, 79, 80
flagellates 155
Foadia 58
Foadiini 58
Ford, Henry 49
forest canopy fauna 76
Fortey, Richard Alan 198
fossils 4, 10, 40, 48, 50–2, 90, 92, 98, 109, 110, 122, 130, 132, 167, 169, 222, 225, 226; *see also* paleontology
Franklin, Benjamin 130
Franklinia alatamaha 131
fruits 95
fungi 29, 30, 32, 77, 79, 80, 94, 100, 101, 132, 154, 162, 163, 170–3, 184, 206
fungus gnat 170–7

Gaffney, Eugene S. 112
Galapagos Islands 22
galaxies 157
Galileo 99
Gelae baen 58; *G. donut* 58; *G. fish* 58; *G. rol* 58
GenBank 4, 35, 82, 177
general reference system 30, 50, 113, 115
general tracks 188
genetics 19, 20, 71, 72, 93, 109, 114, 147, 200, 208, 210, 214, 226; molecular 3, 5, 20, 50, 71, 94, 100, 141, 142, 148, 165, 214, 228; population 20, 71, 78, 94, 123, 140, 141, 200, 214

genitalia 55, 58
genome 5, 32, 50, 77, 92, 94, 101, 110, 147, 174, 177
geology 11, 50, 52, 100, 109; events 52, 87, 108; 188; historic record 78, 132, 195, 203; time 36, 37, 78
Gettysburg 53
Gibbon, Edward 102
Gifford Pinchot National Forest 163
Ginkgo biloba 122
Ginkgoaceae 122
Gnamptogenys pleurodont 79
Goldie Locks zone 104
Gondwana 89
gorilla 93, 101
Gould, Stephen Jay 88
Great Basin National Park 132
Great Smoky Mountains National Park 144
Grimaldi, David A. 76
Guatemala 85

Half-Earth Project 1, 25, 40, 41, 118–20, 123, 134, 161, 199, 218
Hamilton, Andrew 56
Heerz lukenatcha 58; *H. tooya* 58
Hennig, Willi 47, 50, 63, 65, 114, 115, 217, 228
herpetologist 32
Hippopotamus amphibius 24
hipposudoric acid 24
Hitchcock, Alfred 40
Hoffmann, Richard 209
Holocene 90, 198, 203, 204, 206
hominids 36, 37, 122; *see also Homo*
Homo erectus 36; *H. floresiensis* 36; *H. habilis* 36; *H. neanderthalensis* 36; *H. sapiens* 36, 38, 39; *see also* humans
homology 5, 50, 95, 100, 145, 146, 148, 176, 209
Horn, George Henry 144
Hortus 209
hot springs 77
hotspots, biodiversity 83, 119, 133, 168, 188
humanities 51, 52, 62, 64
humanity 12, 16, 22, 31, 35, 38, 39, 45, 54, 59, 135, 155, 157, 168, 204, 230
humans 16, 21, 31, 36–9, 41, 42, 52, 83, 93, 101, 108, 110–12, 129, 132, 134, 149, 152, 153, 157, 160, 161, 203, 207, 228; *see also Homo sapiens*
hydrosphere 77
Hylecoetus 53; *H. dermestoides* 53
hypotheses 5, 9, 11, 15, 20, 26, 28, 30, 35, 44, 45, 57, 63–9, 71, 77, 82, 87, 96, 100, 105, 109, 112, 113, 141, 145, 147, 148, 168, 173, 176, 177, 182, 195, 200, 226, 230, 231

ichthyologist 32
idea space 53
identification, of species 3–5, 20, 30, 61, 67, 71, 82, 93, 99, 100, 106, 109, 142, 147, 165, 169, 172, 175, 178, 183, 190, 194, 227, 229; crowd-sourced 173, 176, 177
imaging, digital 69, 100, 168, 173–5, 190, 210, 211
Indigenous people 76, 193
Indonesia 83, 185
information science 19, 50, 187, 194, 223
Ingersoll, Robert G. 39
insects 9, 29, 30, 36, 51, 56, 58, 61, 65, 75–7, 79, 80, 99, 100, 102, 110, 113, 133, 134, 144, 154–6, 170, 171, 189, 191, 200, 201, 206, 216; abundance 133
intellectual, freedom 165, 176, 187, 194; intimacy 53; property rights 193
interior design 156
inventory, taxonomic 6, 10, 16–18, 23, 25, 27, 31, 35, 42, 65, 69, 74, 76, 78, 80, 100, 106, 104–16, 118–20, 124, 127, 130, 132, 134, 135, 149, 152, 157, 163, 168, 172, 181–96, 199, 201, 203, 206, 207, 210–12, 215, 216, 218, 220, 221, 223, 225, 227–9, 231
invertebrates 132, 133, 152, 184, 185, 192
IPNI (International Plant Names Index) 191
Ipomoea 169
Irpex lacteus 94

Jamaica 56
Janzen, Daniel H. 8
Jetz, Walter 185
Judd, Walter Stephen xii
Jurassic 89, 203

Kennedy, John F. 183
Keroplatidae 170
Kierkegaard, Søren 141
Kimmerer, Robin 157
Kling, Matthew M. 122

knowledge-base 106, 167, 172, 174, 176, 188
Kolbert, Elizabeth 129, 152
Kotter, John 97
Krulwich, Robert 56
K-selection 79
Kudo, Richard R. 155

Lacrymaria olor 155
Lady Gaga 218
landscape architecture 149
language 25, 31, 53, 55, 62, 71, 90, 111–13, 162, 220; history 52
Large Hadron collider 31
Lascaux 98
laws, evolution 71, 89, 101; collecting 192, 194; physics 29, 32, 71
Lebdioui, Amir 223
Leiodidae 56
Lent, Herman 144
Leonardo 54
Lepidoptera 50, 133
lepidopterist 32
Lepidotrichidae 110
lichenologist 32
life list 175
Lincoln, Abraham 39
Lindroth, Carl 57
Linnaeus 11, 28, 30, 31, 53, 65, 75, 95, 100, 112, 135, 144, 179, 181, 183, 193, 206, 217, 225–8
Livingston, David 205
London 9, 54, 90, 95, 146, 189, 199, 200–3, 206, 216, 218
lotus 162
Louca, Stilianos xii
Louv, Richard 157
Louvre 36
Lovejoy, Thomas E. 119
LTER (Long-Term Ecological Research Network) 226
Luc, Michel 63
Lucid, keys 67
Lycoperdina ferruginea 131
Lymexylidae 53, 55, 56

Mace, Georgina 124
Macleay, W.S. 191
Madagascar 132, 135, 185
maize 83
malacologist 32
malaria 83
mammals 35, 80, 132
mammary glands 15, 28
manuscripts, history of 52
Mars 31, 34, 104, 107, 108, 230
Marsh, Paul M. 58
Marx, Groucho 80
mass extinction *see* extinction
May, Robert 167
Mayr, Ernst 140
McCabe, Timothy 53
McHugh, Joseph V. 56, 135
Meier, Rudolf 86
Meierotto, Sarah 82
Melanoplus spretus 133
Melittomma 55
Mendel, Gregor 113
metabarcoding 3
metagenomics 77, 87, 98, 105
metaphysics 46, 52
Mexico 55, 95, 145
Meyrick, Edward 50
Michelangelo 54
microbes 1, 4, 15, 17, 27, 30, 32, 80, 83, 105, 169, 206, 226
microbiologist 32
microscopy 100, 147; *see also* telemicroscopy
Mihi itch 191
Miller, Kelly B. 56, 58, 144, 218
Miller, Scott E. 130
Milman, Oliver 134
Minelli, Alessandro 95
Mishler, Brent D. 46, 47, 122
mites 5, 30, 32, 51, 79, 80, 184
molecular clock 52, 88
monarch butterfly 133
monograph, taxonomic 4, 10, 44, 70, 76, 101, 143–5, 149, 163, 168, 169, 174–7, 187, 188, 195, 218, 226
mononomials 112, 113
monophyletic groups 14, 145, 146
Monotropa uniflora 132
moonshot 1, 183, 226
morphology 3–6, 9, 10, 19, 40, 44, 47, 48, 50, 51, 53, 55, 62, 68, 69, 71, 83, 89, 91–102, 106, 110, 139, 142, 143, 146, 147, 155, 162, 165, 169, 173, 192, 199, 211, 225, 226, 229
morphospecies 108
Morrison, Jim 203
mosquito 83, 162
Moura, Mario R. 185
mouse 192
Muir, Frederick 102

Muir, John 156
Musca domestica 28
Museum National d'Histoire Naturelle, Paris 49, 53, 190, 201
Myanmar 83
Mycetophilidae 170, 174
mycologist 32
myriapods 95
myrmecologist 183
Myxomycetes 51, 56, 154, 189, 215, 218

names, celebrity 217; humorous 58; information content 29; naughty 58; scientific 8, 9, 11, 14, 19, 20, 26, 28, 29, 32, 35, 43, 45, 57–9, 63, 71, 75, 83, 106, 111, 112, 141, 144, 165, 169, 173–6, 182, 188, 190, 191, 199, 205, 217, 225, 228, 230; vernacular or common 8–9, 76, 85, 99
NASA (National Aeronautics and Space Administration) 82, 183
National Academy of Sciences, US 186
National Cancer Institute 163
National Geographic 51
National Institutes of Health 62, 185, 186
National Museum of Ireland 205
National Science Foundation 10, 62, 107, 148, 187
natural history 32, 48, 50, 77, 83, 95, 98, 142, 143, 146, 157, 162, 167, 169, 173, 176, 206, 215, 225
Natural History Museum, Dublin 205; London 9, 90, 146, 189, 199, 200–3, 206, 218
natural history museums 3, 8, 9, 11, 17, 19, 21, 23, 31, 42, 44, 52, 53, 62, 66, 67, 69, 71, 72, 78–80, 84, 85, 87, 88, 99, 105, 106, 108, 109, 114, 121, 134, 140, 149, 155, 168, 174, 182, 187, 198, 201, 203–5, 207, 209–12, 217, 218, 225, 231; exhibitions 177, 188, 206, 211
natural resources 8, 12, 41, 89, 111, 130, 149, 161, 221, 222
natural selection 16, 22, 28, 32, 37, 47, 78, 88, 94, 162, 203, 221, 222
nature, processes 3, 10, 14, 26, 34, 36, 56, 92, 94, 96, 100, 107–9, 114, 141, 221, 225
neanderthals 23, 36, 37
Nelson, Gareth 47, 98, 99, 217
nematodes 32, 77, 80, 184
nematologist 32
Neoceroplatus betaryiensis 170
NEON (National Ecological Observatory Network) 226
neptune 230
neurobiology 62
New Jersey Pine Barrens 53
New Systematics 20, 66, 71, 140
New York Botanical Garden 190; State Museum 53
New York Times 80
New Zealand 191
Newton, Isaac 53
NIH *see* National Institutes of Health
Nixon, Kevin C. 66
Nops 58
norhipposudoric acid 24
North America 57, 83, 89, 144, 193
Notnops 58
NSF *see* National Science Foundation

Occam's razor 145
odonatologist 32
Ohio State University 48, 49, 61, 178, 216
Olea europaea 83
olive tree 83
ontogeny 5, 10, 30, 32, 40, 48, 51, 65, 70, 93–96, 98, 100, 101, 147, 165, 167, 169, 177, 225
Open Linnaean Academy 189
Orchidaceae 75
orchidologist 32
Organization of American States 193
ornithologist 32
Orthohalarachne attenuata 131
Owen, Richard 70, 146
Owl, spotted *see Strix*
Oxford University 146, 204
Ozark Mountains 145

Pacific Northwest 145
Pacific yew *see* Taxus
Pakaluk, James 58
paleontology 30, 52, 93, 98, 101, 121, 142, 165, 177, 182, 229; *see also* fossils
Panama 108, 144
Panthera onca 79
parallelism, three-part 98
Paramecium caudatum 51
Paranthropus robustus 36
Parastratiosphecomyia stratiosphecomyioides 112
Pariat, Janice 225
Paris 204
Parliament 146

passenger pigeon 130
Pearce, Mick 162
Penicillium roqueforti 37
Permian 221
Peru, Lima 56; Rio Tambopata 57
phenetics 11, 141
Philip, Cornelius 58
philosophy of science 52
Phobaeticus chani 9
phycologist 32
phylocode 112, 113, 182
phylogenetic thinking 66; tree 14, 75, 86
phylogenetics 3, 142
phylogeny 30, 40, 64, 68, 74, 75, 88, 89, 98, 99, 101, 110, 114, 115, 169, 175, 176, 182, 195, 209
Phytotaxa 149
Pieza deresistans 58; *P. kake* 58; *P. pi* 58; *P. rhea* 58
Pinker, Steven 112
Pinus 8
Planetary Biodiversity Inventory projects 187
plants 3, 9, 15, 17, 30, 32, 42, 51, 68, 83, 108, 130, 134, 158, 204, 206, 219; beauty 51; cancer cure 163; cultivated 209; endemism 132; extinction 129, 133, 156; flowering 95, 156, 184, 191; garden 152; medicinal 49; nomenclature 144; number 75, 76, 192; registry names 191; rare 154; showy 110; survey 183–184; vascular 28
PlantSnap 67
Platnick, Norman 16, 49, 58, 84, 86
Pleistocene 203
pluralism, of species concepts 85
Popper, Karl 64
Popular Science 80
population genetics 78
population thinking 66, 71, 140
primary production 160
problem solving 84
prokaryotes 8
protists 38, 51, 155
pteridologist 32
Ptolemy 63
public awareness 62
puffball 79, 131
Pyron, Alexander 37

quadrupeds 65
Quercus alba 112; *Q. rubra* 35
Quicke, Donald 218

rainforest 9, 14, 76, 83, 85, 108, 154
Raphael 54
Raven, Peter H. 130, 134, 161, 183, 184, 217
Ray, John 84
reciprocal illumination 5
reductionism 71, 72
Reid, John W. 119
reptiles 22, 31, 32, 65, 79, 132, 185
resources, non-renewable 8, 23, 160, 161
revelation, of phylogeny 98
revision, taxonomic 44, 70, 141, 143–6, 148, 163, 168, 169, 174, 187, 226
Riga toni 58
right questions in history 77
robotics 175, 177, 210
Rome 88
Roosevelt, Theodore 153
Rosaceae 15, 19
Rosen, Donn Eric 85
Rosetta stone 168
Ross, Herbert H. 140
Roswell greys 104
rotenone 154
round worms 77, 184; *see also* nematodes
royal garden of medicinal plants, Paris 49
Royal Society 186
Royal Society Te Aparangi 181
r-selection 79
Ruse, Michael 100
Russell, Kimberly N. 178
Russulaceae 132
Ryan, Catherine 156

Sagan, Carl 41, 108, 123
Sagorny, Christina 190
Saint Sophia 102
Salt, George 79
Sandero Luminoso 57
São Paulo 207
sarcodines 155
Saturn 230
Saunders, Thomas E. 71
Scala naturae 38
scanning electron microscopy 100
scholarship, chain of 45
Schuh, Randall T. 11
Schultz, Colin 102
science 4, 6, 9–12, 14–21, 29, 36, 40, 41, 43–6, 49, 52, 61–5, 69–71, 75, 76, 78, 84, 87, 88, 93, 94, 105–7, 113, 114, 121, 123, 135, 139, 140, 156, 157, 165, 169, 176, 182, 192, 200, 201, 205, 207, 210,

212; citizen 211; *con amore* 54; popular 140, 143, 149, 187, 188, 208, 214; settled 47; *vs.* technology 146
scientific method 52
Scydmaeninae 29
sea lion 22, 131
Sea World 131
seal 131
Seberg, Ole 148
self-promotion 214–219
SETI (Search for Extraterrestrial Intelligence) 104
sexes 5
Shakespeare, William 97
shared-derived character 11, 65, 78, 101, 114; *see also* synapomorphy
Sharkey, Michael 6, 142
Sharp, David 102
silverfish 110
slime mold *see* myxomycetes
Slipinski, Adam 58
Smithsonian Institution 24, 76, 201, 202, 209
Snodgrass, Robert 100
social sciences 62
Solem, Alan 58
South America 54, 56, 57, 89, 170
South Korea 223
spacetime continuum 53
Spangler, Paul J. 58
speciation 52, 66, 67, 74, 87, 88, 108, 114, 141
species 3, 8, 14, 66, 84; arbitrary 65, 66; beauty 51; biological 85; biomimetic models 23, 111, 160; charismatic 120; complexity 51; conservation 34, 36, 40, 110, 118; definition 84, 87; description 4, 44, 83; discovery 5, 26, 29, 75, 119, 140, 167, 181, 218; distribution 89, 132; diversity 36, 118; economy 220; ecosystems 34, 107; evolution 110; exploration 42, 165, 193; extinction 37, 92, 129, 146; hall of fame 217; human connections 152; identification 30, 61, 69, 142, 165; in-the-making 141; invasive 107; inventory 31, 104, 106, 181; language 112; level 94; love of 46, 54; naming 35, 58, 106, 113, 217; nonessential 35; nonsense 66; number 75, 76; sibling 5; specialization 32; systematics 30; top ten 216; uses 28, 35; value 36; works of art, cf. 36
species-scape 74, 75, 79
specimens 53
SPIDA-Web 68
spiders 66, 90
spinneret 66
Spock 45
springtails 77, 79
spruce-fir forest 135
Staphylinidae 29
stars 157
Stenocara gracilipes 220
statistics 52, 64
stratigraphy 52, 98
Strix occidentalis 110
struggle for survival 22, 38
Stygobromus albapinus 132
subspecies 57, 66
Suez Canal University, Ismailia 47
super-society 186
Supreme Court, US 188
sustainability 12, 14, 21, 23, 24, 34, 42, 84, 91, 135, 148, 158, 160–3, 202, 220–3
Swordtail fish 85
synapomorphy 101, 114, 145, 229; *see also* shared-derived character
synonyms 28
systematic biology *see* systematics; taxonomy
Systematic Zoology 47
systematics 9, 11, 14, 29, 30, 41, 44; achievements 28; aim 94; amateur 69; collections 167, 196; *con amore* 54; conservation 119; contemporary 62; cosmology 26; definition 9–10; descriptive 62, 67; devalued 19, 20; economics 220; education 190; enigma 44; goals 87, 148; history 52, 64; hypotheses 64; independence 71, 109; intellectual excitement of 55; intersectionality 44, 50; language 112; marginalization 41; merely descriptive 67; minimalist 6; mission 30; misunderstood 63; molecular 3, 10, 68, 93, 147; morphological 92; neglect 42, 135, 147; passion 54; perspective 26; phylogenetic 63, 114; promotion 214; reimagined 31, 43, 165, 169; reputation 63; revival 184; scope 26; service 67; success 4; support 84; theoretical 50, 65; thought traditions 6, 44, 157; under siege 139; virtues 48; *see also* taxonomy

Systematics Agenda 2000 186
SWAT teams 171
Swimsuit, sharkskin 162

Tabanus nippentuck 58; *T. rhisonshine* 58
Taintnops 58
Tambopata, Rio 57
Taraxacum officinale 8
tardigradologist 32
Taxa, higher 14
Taxol *see Taxus*
taxonomic impediment 145, 191
taxonomic renaissance 1, 6, 12, 42, 95, 120, 134, 141, 167–80, 191, 210, 221, 223, 225
taxonomy 3, 5, 6, 11, 44–60, 113, 132, 139, 140, 145, 148, 187; amateur 54, 69; applied 26, 28, 29, 99; bad 31, 57; biomimicry 24, 90, 111, 162; chain of scholarship 53; citations 144, 149; collections 70, 140, 177, 200; conservation 124; curricula 143, 190; cyber-enabled 19, 187, 189, 210, 211; decimation of 9; definition 9–10; descriptive 23, 28, 46, 69, 78, 89, 90, 99, 163, 182, 185, 223, 225, 229; ecology 108, 182; funding 186; genetics 71, 78, 93, 113; goals 19, 82, 96, 97, 141, 148, 167; Hennigian revolution 47, 50; humanities 52; hypotheses 77; independence 20, 64, 139; information content 10, 19, 68, 69, 223; intellectual freedom 188; intellectual rewards 11, 55; Linnaean 226; maximalist 82; minimalist 6, 78, 82; misconceptions about 65, 140; mission 30, 169; modern 14; modernizing 165; molecular-based 27, 68, 82, 95–9, 114, 142, 148, 182, 226, 229; morphology-based 9, 101; museums 177, 209; natural history 157; neglect 115, 140; observational 64; prestige 11, 43, 63, 139, 191, 214; promotion 214–19; quality of evidence 28; redefining 93, 94; reimagined 167–9, 198; service 20, 64, 68; specialization 32; subjectivity 63; synonymy with systematics 11; tasks 50, 156; teamwork 172; *see also* systematics
Taxus brevifolia 163; *T. floridana* 163
teamwork 4, 45, 50, 70, 165, 170–2, 176, 181, 187, 189, 192, 194, 207, 223
technology 19, 62, 146, 165; overemphasis 12, 20
tectonics, plate 89
telemicroscopy 69, 170, 174, 193
termites 162
testability 46, 64, 77
Thailand 218
theology 46
Tisentnops 58
Titanic 93, 120
tokogenetic relationships 114
traits 57, 66, 67, 74, 77, 78, 114
tree of life 88
tree-hugger *ethos* 121
Triatominae 144
Triceratops 207
Tricholepidion gertschi 110
tulips 124
Twain, Mark 96
type specimens 170, 175, 190; e-types 175, 195

United Nations 186
universe 15, 27, 29, 31, 32, 42, 52, 104, 105, 108, 157, 158, 199, 229

Van der Valk, Tom 92
Vanilla 29
Velcro 162
Venter, Craig 87
vertebrates 65
Vicariance *see* biogeography
Volvox globator 51
Vorticella campanula 51
V-2 rocket 105

Wallace, Alfred Russel 70
walrus 131
Washington Post 37
Waterhouse, Alfred 146
Web of life 35
Webb telescope 31
Wheeler, Olivia 53
Whitman, Walt 52
Wickham, Henry 58
Wilde, Oscar 6, 219
Williams, David M. xii
Wilson, E. O. 1, 25, 31, 33, 36, 40, 68, 76, 79, 80, 183
Wilson, John James 142
Windshield anecdote 133
Winsor, Mary P. 44, 71
Wisdom 124

Wood, John R. I. 169
World Intellectual Property Organization 193
Wygodzinski, Pedro 110, 144

Xenophobia 62
Xiphophorus 85; *X. montezumae* 86

Ytu brutus 58

Zea mays 83, 217
Ziegler, Alexander 190
Zimbabwe 162
Zoo 24, 93, 98, 120, 205
ZooBank 169, 174, 191
Zootaxa 149